Success guides

Leckie × leckie
publishers

D0532649

Intermediate 1
Biology

× George Milne ×

Contents

Unit 1 – Health and Technology

Unit 2 – Biotechnological Industries

Unit 3 – Growing Plants

Exam skills and support

How to use this revision guide

Congratulations on deciding to use this Int 1 course guide book. The book is designed to **build your confidence** and enable you to **learn** and **remember** the important course information so that you can **successfully answer all the questions** in the end of unit tests and the final exam.

In each section of the book:

- The **key words** have been typed in **'bold'** to remind you that these are the words that the **examiner wants to see in your answers**.

- The **Quick Tests** at the end of each section will help you to **check that you have learned the main points**.

- A lot of the information that you are required to know has been displayed in the form of **'Mind Maps'**. A Mind map is a **type of diagram** that **displays information** in a way that makes it **easier to remember** when sitting your End-of-Unit (NAB) Tests and the Final Exam.

Appendix 1 gives you some examples of the types of questions that you will be asked in the final exam. It also gives you guidance on how to draw **line graphs**, **pie charts** and **bar charts** so that you will gain maximum marks for these activities.

Some of the questions involve **carrying out calculations**. **Appendix 3** will help you to understand clearly how to work out **percentages**, **averages** and **ratios**.

The guide also highlights **the Practical (LO2) Activities** that you have to successfully carry out as part of the **course internal assessment** (i.e. the part of the course that is assessed in school).

Investigations are also part of the internal assessment. The guide suggests areas of topics where you may decide to carry out an investigation. **Appendix 2** gives you clear instructions on how to **plan**, **carry out** and **write a report** about an investigation.

The **Glossary** at the end of the book describes each of the important terms that you need to know.

Assessment Summary

Internal Assessment

This occurs **throughout the course** and involves:

- Passing **NAB (National Assessment Bank) Tests** at the end of each unit of study.

- Passing the **LO2 Practical Activities** (Your teacher will observe you as you carry out each of these activities).

- Planning, carrying out and writing a report about an **investigation**.

External Assessment

This involves sitting an **exam** at the **end of the course**. The exam consists of **one paper** of 1 hour 30 minutes. **Section A** contains 25 **multiple choice** questions. **Section B** contains **structured** questions. Both sections contain questions that test your **knowledge** of the topics which you have studied and questions that test your ability to **solve problems** and your **understanding** of the LO2 **practical work** you have carried out.

Top Tip

Get hold of a copy of Leckie and Leckie's collection of Official SQA past papers so that you can see and practise the type of questions you are going to face in the final exam.

The meaning of health

The health triangle

There are **three** main aspects of health: **physical**, **mental** and **social**. These are illustrated in the form of a **triangle**.

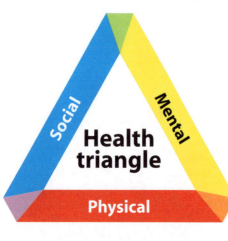

Physical health

Good physical health involves:

- eating a healthy diet
- taking regular exercise
- avoiding unnecessary health risks, e.g. smoking, drinking too much alcohol, taking drugs.

Mental health

Poor mental health can be **caused** by:

- worry
- stress
- low self-esteem
- lack of self-confidence.

Mental health problems can be **overcome** by:

- asking for help
- talking things over with friends
- finding time to relax
- improving your level of organisation.

Social health

Good social health involves:

- communicating well with other people
- enjoying activities with family and friends.

When things get out of balance

The **triangle collapses** if any one of the sides is removed.

Our overall health will be affected if something goes wrong with any of the sides of the health triangle.

Top Tip
Make sure you know the sides of the health triangle and can describe the three main aspects of good health.

Quick Test

1. Name the **three** sides of the health triangle relating to good health.

2. List **three** causes of poor mental health.

3. Give **two** examples of ways in which mental health problems can be overcome.

4. List **three** ways of improving physical health.

5. Describe how people keep socially healthy.

Answers 1. Mental, physical, social. **2.** Worry, stress, low self-esteem, lack of self-confidence. **3.** Asking for help, talking things over with friends, finding time to relax, improving your level of organisation. **4.** Eating a healthy diet, taking regular exercise, avoiding unnecessary health risks e.g. smoking, drinking too much alcohol, taking drugs etc. **5.** They communicate well with other people and enjoy activities with family and friends.

Physiological measurements

Why do we take physiological measurements?

Special instruments are used to take **physiological measurements** such as **heart rate**, **blood pressure** and **body temperature**.

When **compared with the average normal values** these measurements can give an **indication of the state of a person's health**.

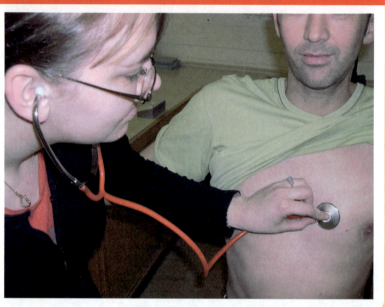

Low and High Tech Instruments

The table below gives examples of Low and High Tech Instruments used to make physiological measurements.

Physiological Measurement	Low Tech Method	High Tech Method
Temperature	Clinical thermometer	Digital thermometer
Blood pressure	Stethoscope and mercury manometer	Digital sphygmomanometer
Heart (pulse) rate	Stethoscope Finger and stopwatch	Pulsometer Heart rate monitor
Percentage body fat	Skinfold callipers	Digital body fat sensor

Advantages and disadvantages of using Low Tech Instruments

Advantages:
Usually **less expensive** than High Tech instruments
Lower maintenance costs.

Disadvantages:
Human errors can occur when using them
Cannot usually be connected to computers.

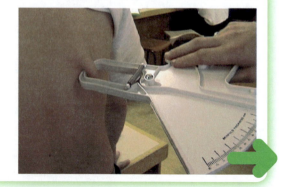

Advantages and disadvantages of using High Tech Instruments

Advantages:

Give faster, more accurate readings

Can usually be connected to computers.

Disadvantages:

More expensive to make and often have high operating costs.

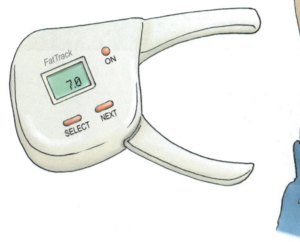

Quick Test

1. Name a piece of equipment that is used to measure body fat.

2. What is a sphygmomanometer used to measure?

3. Give **two** ways in which heart rate can be measured.

4. State **two** advantages of using high tech equipment when making physiological measurements.

Answers 1. Skinfold callipers, digital body fat sensor. **2.** Blood pressure **3.** Using a stethoscope, pulsometer or heart rate monitor. **4.** Give faster, more accurate readings, can usually be connected to computers.

The heart and blood vessels

The function of the heart

The **circulatory system** is made up of a **muscular pump** (the **heart**) and inter-connecting tubes (**blood vessels**) that carry blood to all parts of the body.

The heart **keeps blood flowing through the blood vessels**.

The structure of the heart

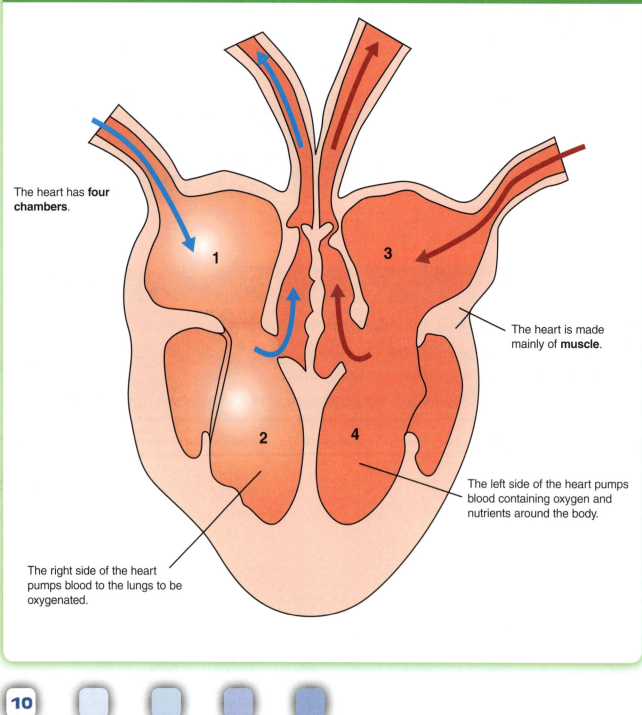

The heart has **four chambers**.

The heart is made mainly of **muscle**.

The left side of the heart pumps blood containing oxygen and nutrients around the body.

The right side of the heart pumps blood to the lungs to be oxygenated.

Blood vessels

Blood is carried around the body in **blood vessels**.

The three main types of blood vessel are **arteries**, **veins** and **capillaries**.

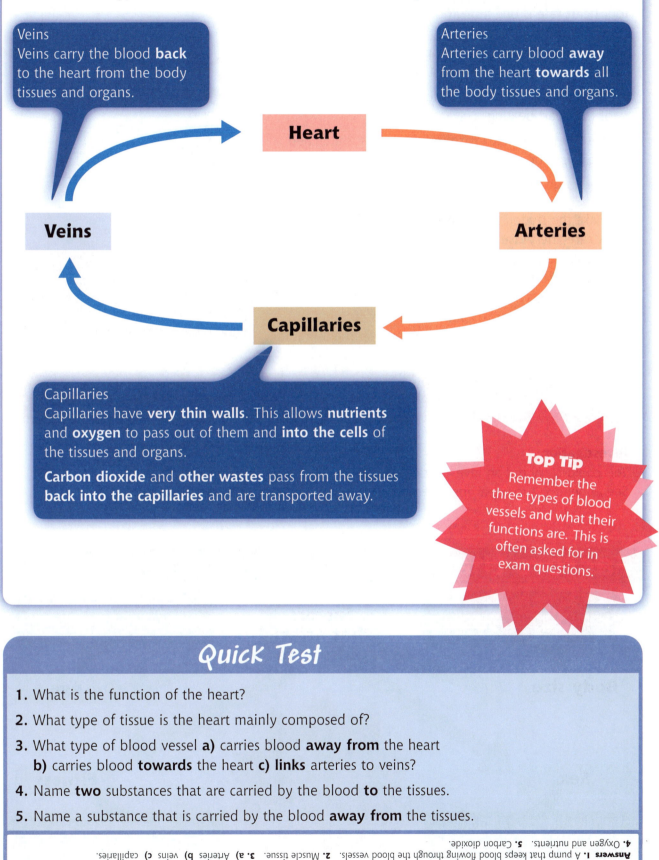

Veins
Veins carry the blood **back** to the heart from the body tissues and organs.

Arteries
Arteries carry blood **away** from the heart **towards** all the body tissues and organs.

Capillaries
Capillaries have **very thin walls**. This allows **nutrients** and **oxygen** to pass out of them and **into the cells** of the tissues and organs.

Carbon dioxide and **other wastes** pass from the tissues **back into the capillaries** and are transported away.

Heart

Veins

Arteries

Capillaries

Top Tip
Remember the three types of blood vessels and what their functions are. This is often asked for in exam questions.

Quick Test

1. What is the function of the heart?

2. What type of tissue is the heart mainly composed of?

3. What type of blood vessel **a)** carries blood **away from** the heart **b)** carries blood **towards** the heart **c) links** arteries to veins?

4. Name **two** substances that are carried by the blood **to** the tissues.

5. Name a substance that is carried by the blood **away from** the tissues.

Answers 1. A pump that keeps blood flowing through the blood vessels. **2.** Muscle tissue. **3. a)** Arteries **b)** veins **c)** capillaries. **4.** Oxygen and nutrients. **5.** Carbon dioxide.

Pulse rate

Measuring pulse rate   LO2 Activity

Your heart rate is one **indication of the state of your health. Pulse rate is the same as heart rate** as they both measure the number of contractions of the heart muscle. Heart rate can be measured in several ways:

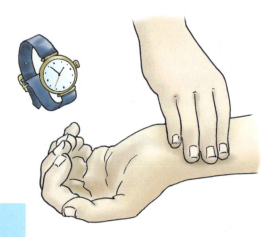

- A **stethoscope** can be used to listen to the sounds of the heart beating.
- **Pulsometer** or **heart rate monitor**. This is attached to a finger or an earlobe.
- **Fingers** on the wrist/neck and a **stop watch**.

1. Make the student **sit quietly** for five minutes.
2. **Locate the pulse** in the wrist/neck correctly.
3. **Count** the numbers of pulses in 15 seconds.
4. **Record** the result in a table.
5. **Repeat** steps 3 and 4 another four times.
6. **Calculate** the average pulse rate in 15 seconds and then multiply by four to give pulse rate in one minute.

Remember:
Repeating readings and calculating an average value **makes the results more reliable**.

To pass this learning outcome you must **successfully measure and calculate the resting pulse rate**.

Most adults have a resting pulse rate in the range 65–74 beats/minute.

Investigation:
You could carry out an investigation on the **effect of exercise on pulse rate**. See appendix 2 for clear instructions on how to write up an investigation report.

Factors that affect pulse rate

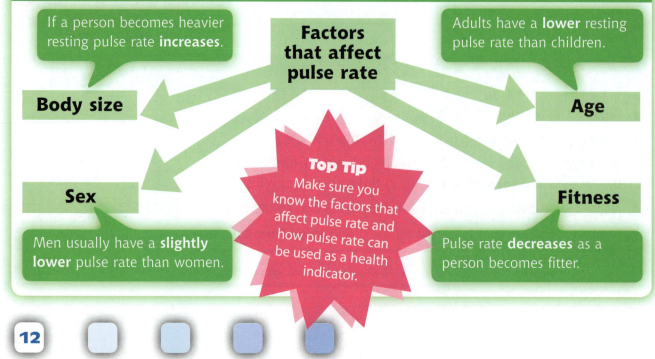

If a person becomes heavier resting pulse rate **increases**.

Factors that affect pulse rate

Adults have a **lower** resting pulse rate than children.

Body size

Age

Top Tip
Make sure you know the factors that affect pulse rate and how pulse rate can be used as a health indicator.

Sex

Fitness

Men usually have a **slightly lower** pulse rate than women.

Pulse rate **decreases** as a person becomes fitter.

Pulse rate as a health indicator

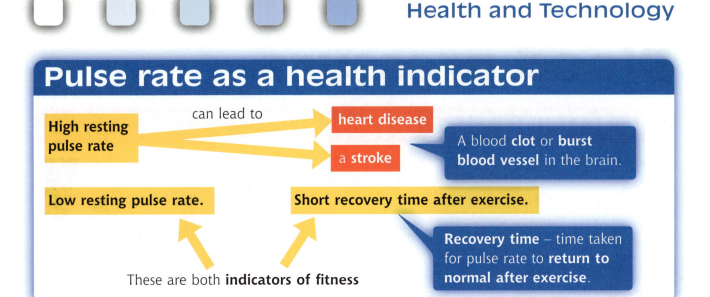

High resting pulse rate

can lead to

heart disease

a stroke

A blood **clot** or **burst blood vessel** in the brain.

Low resting pulse rate.

Short recovery time after exercise.

These are both **indicators of fitness**

Recovery time – time taken for pulse rate to **return to normal after exercise**.

Effect of exercise on recovery time

Recovery time and resting pulse rate can be reduced by taking regular exercise.

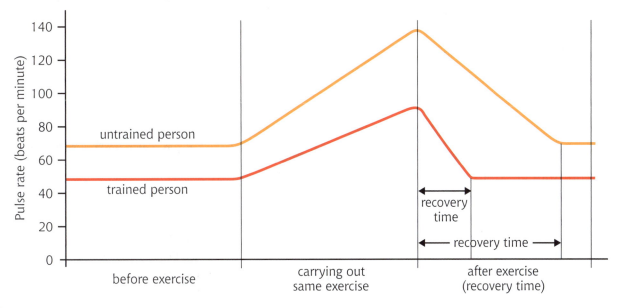

Quick Test

1. Name **two** factors that can affect the **normal range of values** for pulse rate.

2. What immediate effect does exercise have on pulse rate?

3. What effect does regular exercise have on a person's **resting** pulse rate?

4. What term is used to describe the time taken for the pulse rate to return to the normal resting level after exercise?

5. What might be the effect of a high resting pulse rate over a long period of time?

6. State **two** indicators of fitness.

Answers 1. Body size, age, sex, fitness. **2.** Increases pulse rate. **3.** Decreases / will be lower with exercise. **4.** Recovery time / period. **5.** Heart disease, stroke. **6.** Low resting heart/pulse rate, short recovery time.

Blood pressure

Measuring blood pressure

Blood pressure is measured using a **digital sphygmomanometer** or a **stethoscope** together with a **mercury manometer**.

The cuff of the sphygmomanometer is **placed around the arm** above the elbow.

The cuff is **inflated**.

As the cuff deflates **two pressure readings are recorded** on the digital meter.

The **higher** of the two pressure readings is **due to the heart beating** and **pumping blood** into the arteries. The **average pressure** is about **120 mm Hg** (millimetres of mercury).

The **lower** pressure reading is the pressure when **the heart is relaxed** and **filling with blood**. The **average pressure reading** is about **80 mm Hg**.

The two values are written as **120/80**. This is the **average** blood pressure in an **adult**.

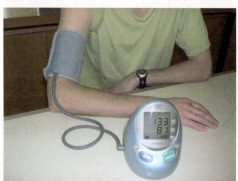

Causes of high blood pressure

Blood pressure **greater than 160/90** indicates **high blood pressure**.

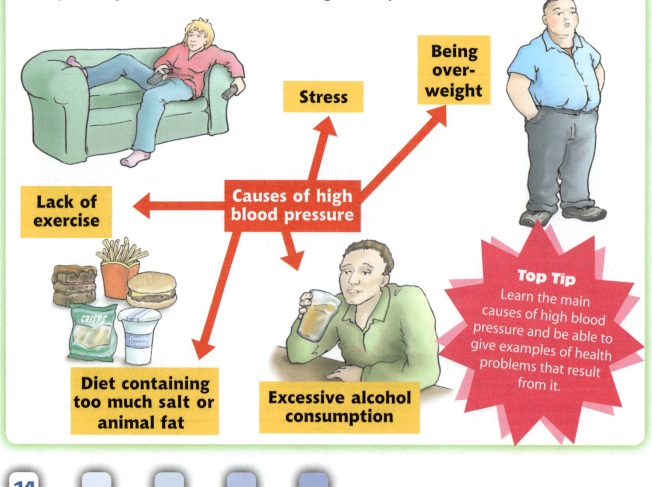

Stress

Being over-weight

Lack of exercise

Causes of high blood pressure

Diet containing too much salt or animal fat

Excessive alcohol consumption

Top Tip
Learn the main causes of high blood pressure and be able to give examples of health problems that result from it.

Effect of high / low blood pressure on health

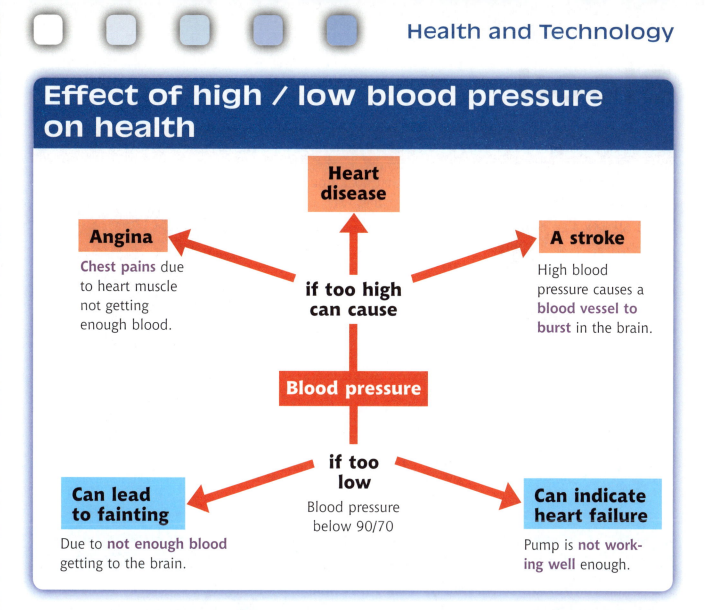

Heart disease

Angina

Chest pains due to heart muscle not getting enough blood.

if too high can cause

A stroke

High blood pressure causes a blood vessel to burst in the brain.

Blood pressure

if too low

Blood pressure below 90/70

Can lead to fainting

Due to not enough blood getting to the brain.

Can indicate heart failure

Pump is not working well enough.

Quick Test

1. List **three** causes of high blood pressure.
2. Describe a person's blood pressure if the readings are greater than 160/90.
3. Name **two** health problems that can result from high blood pressure.
4. What can low blood pressure indicate about the health of a person?

Answers 1. Being over weight, lack of exercise, stress, diet containing too much animal fat / salt, excessive alcohol consumption. **2.** High blood pressure. **3.** Can cause angina, heart disease, a stroke. **4.** Can indicate heart failure.

Blood

The composition of blood

Blood is made up of **red** and **white cells** that float around in a watery fluid called **plasma**.

Red blood cells carry oxygen from the lungs to the body tissues.

Plasma consists of water containing **dissolved substances**, e.g. **sugar**, **salts**, **antibodies**.

White blood cells make chemicals called **antibodies** which **fight infections** in the body.

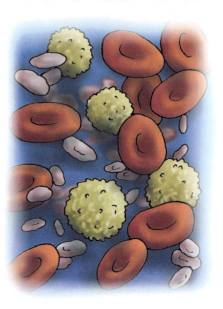

Blood tests and cell counts

A person's **health can be checked** by **testing a sample of their blood** for:

- **Numbers** of the different types of cells
- **Levels** of antibodies and other chemicals.

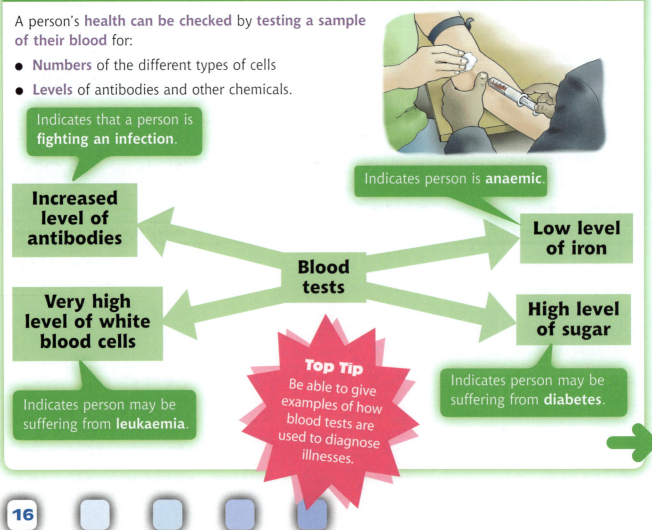

Indicates that a person is **fighting an infection**.

Indicates person is **anaemic**.

Increased level of antibodies

Low level of iron

Blood tests

Very high level of white blood cells

High level of sugar

Indicates person may be suffering from **leukaemia**.

Top Tip
Be able to give examples of how blood tests are used to diagnose illnesses.

Indicates person may be suffering from **diabetes**.

Testing blood for alcohol and drugs

If a drunk driver fails a breathalyser test, a blood test will be carried out to more accurately **determine the alcohol content of their blood**.

Blood tests can also be carried out to test for the presence of drugs, e.g. on athletes to find out if they have been using **illegal drugs** to enhance their performance.

Blood groups

Blood exists in **four main types** or groups – **A**, **B**, **AB** and **O**.

Blood can also be either **Rhesus positive (Rh+)** or **Rhesus negative (Rh–)**.

A patient must be tested to find out what their blood group is before a **transfusion** can be made.

Blood group	Percentage in UK population
A	39
B	15
AB	3
O	43

It is important to make sure that the blood group of the patient and the donor **match** (i.e. **are the same**).

If the wrong blood group is transfused, red blood cells may **clump together** and **block the blood vessels.** This could cause **serious damage to the tissues** and **death of the patient**.

This is the **transfer of blood** from one person to another. This may be necessary if the patient has **lost a lot of blood** during an **operation** or as a result of an **accident**.

Quick Test

1. List **three** components of the blood.

2. What would a **higher** than normal level of antibodies in a blood sample indicate?

3. Name the condition a person would be suffering from if their blood test showed

   a) a **low** level of iron **b)** a **high** level of sugar.

4. Name the **four** blood groups.

5. Which is the least common blood group in the British Isles?

Answers 1. Red cells, white cells, plasma. **2.** Person has an infection. **3. a)** Anaemia **b)** Diabetes. **4.** A, B, AB, O. **5.** AB.

The lungs and breathing

The function of the lungs

The lungs allow **oxygen** in the air to pass **into the blood**. **Carbon dioxide** passes **from the blood** into the air in the air sacs, i.e. the lungs are the organs of **gas exchange**.

The structure of the breathing system

Windpipe: tube through which air passes from the back of the throat towards the lungs.

Bronchi: branches of the windpipe that take air towards each lung.

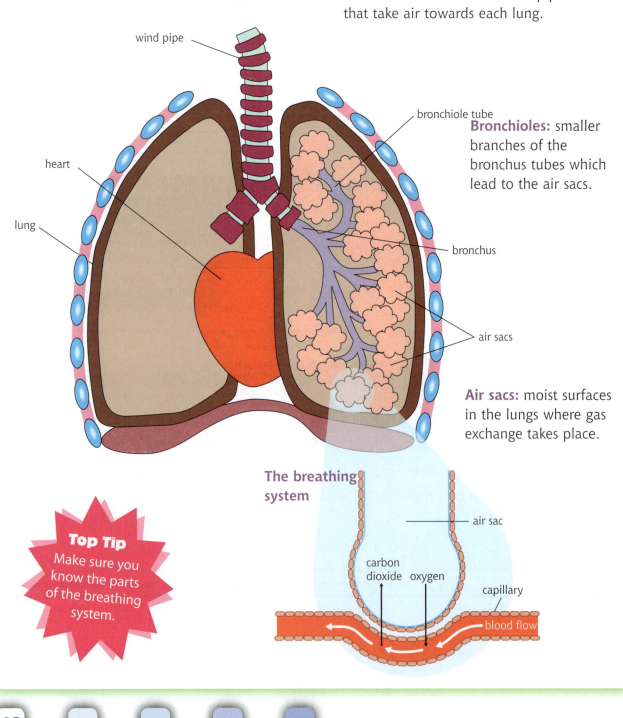

wind pipe

bronchiole tube

Bronchioles: smaller branches of the bronchus tubes which lead to the air sacs.

heart

lung

bronchus

air sacs

Air sacs: moist surfaces in the lungs where gas exchange takes place.

The breathing system

air sac

carbon dioxide   oxygen

capillary

blood flow

Top Tip
Make sure you know the parts of the breathing system.

The breathing rate

The **breathing rate** is the **number of breaths per minute**.

Each **breathing cycle** involves **inspiration (breathing in)** and **expiration (breathing out)**.

Effect of exercise on breathing rate

During exercise breathing becomes **deeper** and **faster**, i.e. **more air is taken in during each breath** and the **number of breaths per minute increases**.

Effects of deeper, faster breathing

More oxygen is taken **into the bloodstream.**

Gas exchange in the lungs **increases.**

More oxygen **reaches the active muscles** so that energy can be released.

Muscles can **continue to function efficiently**.

The **time** required, after exercise, for the rate and depth of breathing **to return to normal resting levels**, is called the **recovery period**.

In a fit person the recovery period is **shorter** than in an unfit person.

Quick Test

1. Describe the route taken by air from the time it enters the breathing system until it reaches the air sacs.
2. Where in the lungs does oxygen pass into the blood?
3. Describe the gas exchange that takes place in the lungs.
4. State **two** effects that exercise has on breathing.
5. How does the recovery time after exercise differ between a fit and an unfit person?

Answers 1. Windpipe > bronchi > bronchiole tubes > air sacs. **2.** In the air sacs. **3.** Oxygen passes from the air sacs into the blood and carbon dioxide passes from the blood into the air sacs. **4.** Breathing becomes deeper and faster. **5.** Recovery time in a fit person is shorter than in an unfit person.

Physiological measurements of the lungs

Tidal volume and vital capacity

Tidal volume is the volume of air breathed in or out of the lungs in one normal breath.

Volume of air in lungs

Time (mins)

Vital capacity is the maximum volume of air that can be breathed out in one breath after a maximum inspiration.

Peak flow

Peak flow is the maximum rate at which air can be forced from the lungs.

Peak flow is used by doctors in the **diagnosis** and **management of patients who suffer from asthma**.

Asthma sufferers cannot force the air out of their lungs as quickly as a healthy person, i.e. their peak flow values will be **lower**.

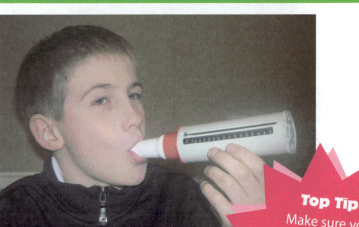

Factors that affect values

The **values of tidal volume, vital capacity** and **peak flow** depend on the **size**, **age**, **sex** and **fitness** of the person being tested.

Top Tip
Make sure you learn the definitions of tidal air, vital capacity and peak flow as these terms are often asked for in exam questions.

Health risks and the effects of smoking

Carbon monoxide in cigarette smoke **reduces the ability of the blood to carry oxygen** around the body.

Tar in cigarette smoke contains chemicals that can cause **cancer**.

Smoking kills

Effects of smoking on health

Less oxygen gets to the body tissues (due to the effect of carbon monoxide) so the **heart has to beat faster** to get the oxygen to where it is required.

If a pregnant mother smokes, **less oxygen and nutrients get from the mother to the developing baby**.

Over a period of time this can put a **strain on the heart** / cause **heart disease**.

Development of baby's tissues may be affected, e.g. brain tissue, during the pregnancy.

Baby's **birth weight lower** than in mothers who do not smoke during pregnancy.

Quick Test

1. Explain what is meant by the term **vital capacity**.
2. Name **three** factors that affect the vital capacity of the lungs
3. What name is given to the volume of air breathed in or out of the lungs in one normal breath?
4. How are **peak flow** values used by doctors?
5. Name an illness that can result from smoking cigarettes.
6. Name a gas in tobacco smoke that reduces the ability of the blood to carry oxygen.
7. What effects may smoking during pregnancy have on the developing baby?

Answers 1. This is the maximum volume of air that can be breathed out after a maximum inspiration. **2.** Size of body, sex, fitness of person. **3.** Tidal volume. **4.** Used in the diagnosis and management of the treatment of asthma sufferers. **5.** Cancer, heart disease. **6.** Carbon monoxide. **7.** Slows down tissue development, reduces birth weight.

A healthy diet

The importance of diet

A **healthy balanced diet** each day is essential for healthy living.
The **main food groups** in a healthy diet are shown below.

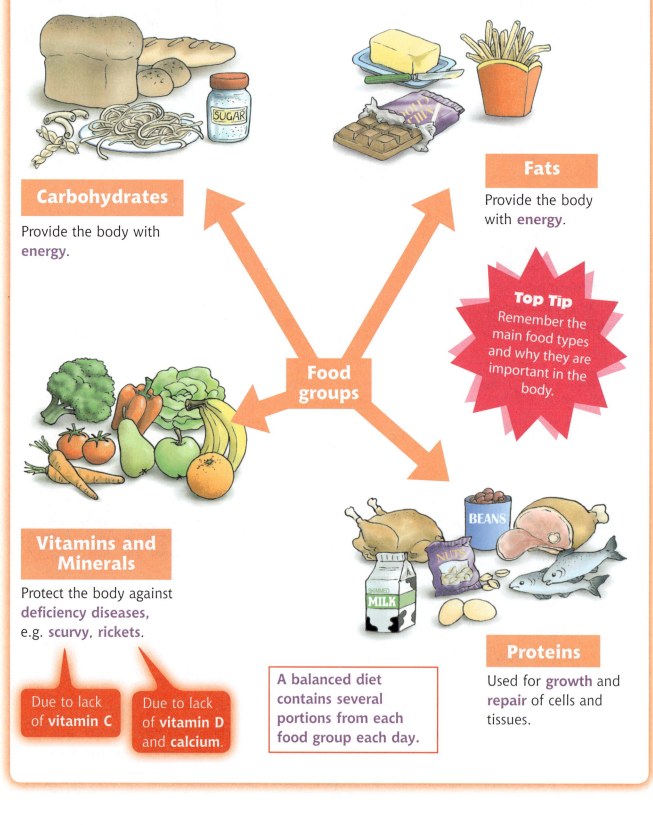

Carbohydrates

Provide the body with **energy**.

Fats

Provide the body with **energy**.

Top Tip
Remember the main food types and why they are important in the body.

Food groups

Vitamins and Minerals

Protect the body against **deficiency diseases**, e.g. **scurvy**, **rickets**.

Due to lack of **vitamin C**

Due to lack of **vitamin D** and **calcium**.

A balanced diet contains several portions from each food group each day.

Proteins

Used for **growth** and **repair** of cells and tissues.

BEANS

NUTS

SKIMMED MILK

Energy requirements

Our body needs energy for:

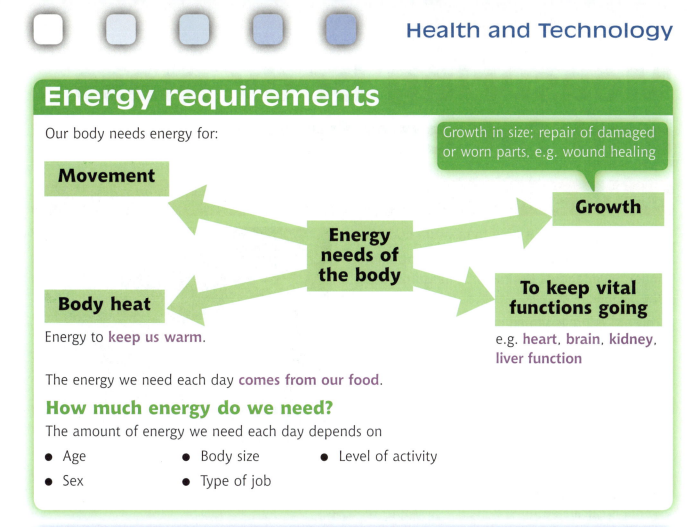

Growth in size; repair of damaged or worn parts, e.g. wound healing

Movement

Growth

Energy needs of the body

Body heat

To keep vital functions going

Energy to **keep us warm**.

e.g. **heart**, **brain**, **kidney**, **liver function**

The energy we need each day **comes from our food**.

How much energy do we need?

The amount of energy we need each day depends on

- Age
- Sex
- Body size
- Type of job
- Level of activity

Energy balance

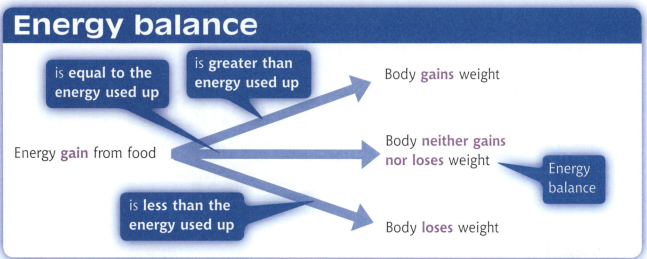

is **equal to the energy used up**

is **greater than energy used up**

Body **gains** weight

Energy **gain** from food

Body **neither gains nor loses** weight

Energy balance

is **less than the energy used up**

Body **loses** weight

Quick Test

1. A healthy diet contains a balance of different food types. Name **three** of the food types.

2. Why are vitamins important in a healthy diet?

3. Which food type is used for body building?

4. List **four** ways in which the body uses energy.

5. If a person's body is in **energy balance**, what is the relationship between the energy gained from food and the energy the body is using up?

Answers 1. Fats, carbohydrates, proteins, vitamins and minerals. **2.** Give protection against deficiency diseases. **3.** Protein. **4.** Movement, body heat, growth, to keep vital functions going e.g. heart, brain, kidney, liver. **5.** The energy gained from food is equal to the energy used up by the body.

Measuring body mass and fat content

Using skin fold callipers

A **skin fold calliper** is used to measure body fat.

A fold of skin is taken between the thumb and index finger and its **thickness is measured** using the calliper.

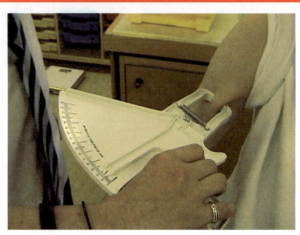

This procedure is carried out at **four positions** in the body:

- front of upper arm
- back of upper arm
- side of waist
- below shoulder blade.

Several readings are taken at each position and an **average value** is calculated.

Remember: Repeating the readings makes the results more reliable.

These **four** skin fold measurements are **added together** and the percentage body fat content of the body is worked out with **reference to a table**.

Factors that affect body mass

The **normal range of body masses (weights)** in a population depends on the following factors:

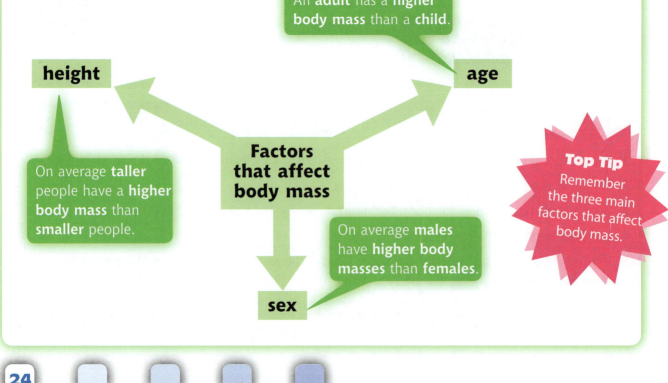

An **adult** has a **higher body mass** than a **child**.

height

age

Factors that affect body mass

On average **taller** people have a **higher body mass** than **smaller** people.

On average **males** have **higher body masses** than **females**.

sex

Top Tip
Remember the three main factors that affect body mass.

24

Implications for health of being overweight / underweight

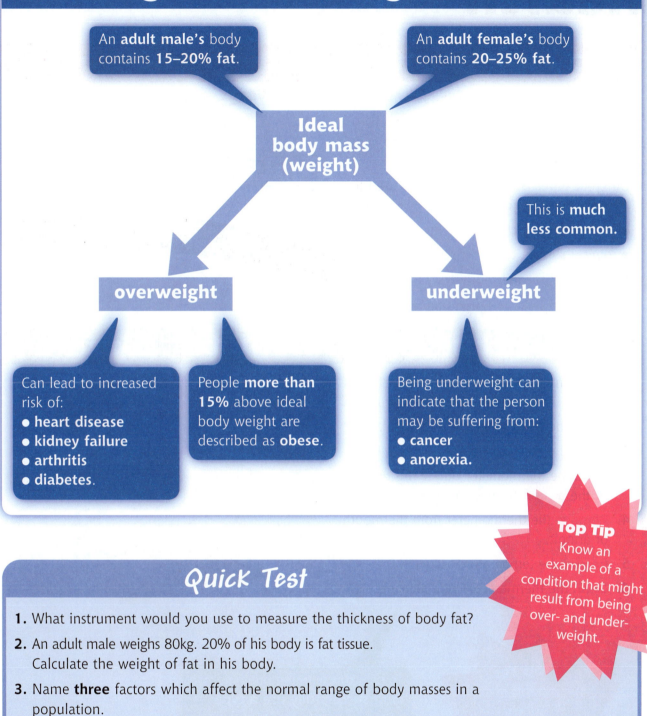

An **adult male's** body contains **15–20% fat**.

An **adult female's** body contains **20–25% fat**.

Ideal body mass (weight)

This is **much less common.**

overweight

underweight

Can lead to increased risk of:
● **heart disease**
● **kidney failure**
● **arthritis**
● **diabetes**.

People **more than 15%** above ideal body weight are described as **obese**.

Being underweight can indicate that the person may be suffering from:
● **cancer**
● **anorexia.**

Top Tip
Know an example of a condition that might result from being over- and under-weight.

Quick Test

1. What instrument would you use to measure the thickness of body fat?

2. An adult male weighs 80kg. 20% of his body is fat tissue. Calculate the weight of fat in his body.

3. Name **three** factors which affect the normal range of body masses in a population.

4. Name a condition that might occur as a result of being obese.

5. Describe one effect of anorexia on the body.

Answers 1. Skin fold callipers. **2.** 16kg. (20% of 80 = 20/100 x 80 = 16kg) **3.** Age, height, sex. **4.** Heart disease, kidney failure, arthritis, diabetes. **5.** Causes severe weight loss.

Body temperature and health

Body temperature

The **normal body temperature** is **37°C**.

Body temperature **must stay close to this value** if the body is to stay healthy and function normally.

Different **types of thermometer** are used to measure body temperature:

- **mercury glass clinical thermometer**
- **digital clinical thermometer**
- **liquid crystal thermometer.**

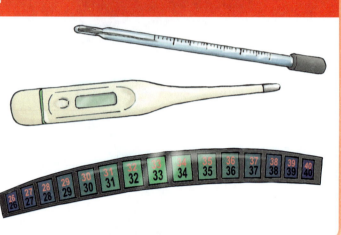

Using a clinical thermometer   LO2 Activity

The steps in the procedure for using a **clinical thermometer** to measure body temperature:

1. **Clean** the thermometer with alcohol.
2. **Reset** the mercury column.
3. **Place** the thermometer **under the tongue** or the **armpit** for two minutes.
4. **Remove** the thermometer from the mouth or armpit.
5. **Record** the temperature accurately.
6. **Clean** the thermometer with alcohol.

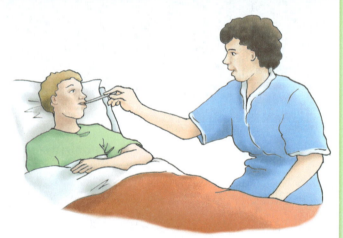

Top Tip
Be able to describe how you would use a thermometer to measure body temperature.

To pass this learning outcome you must **successfully use a type of thermometer to measure a person's body temperature.**

Effects of high temperature on the health of the body

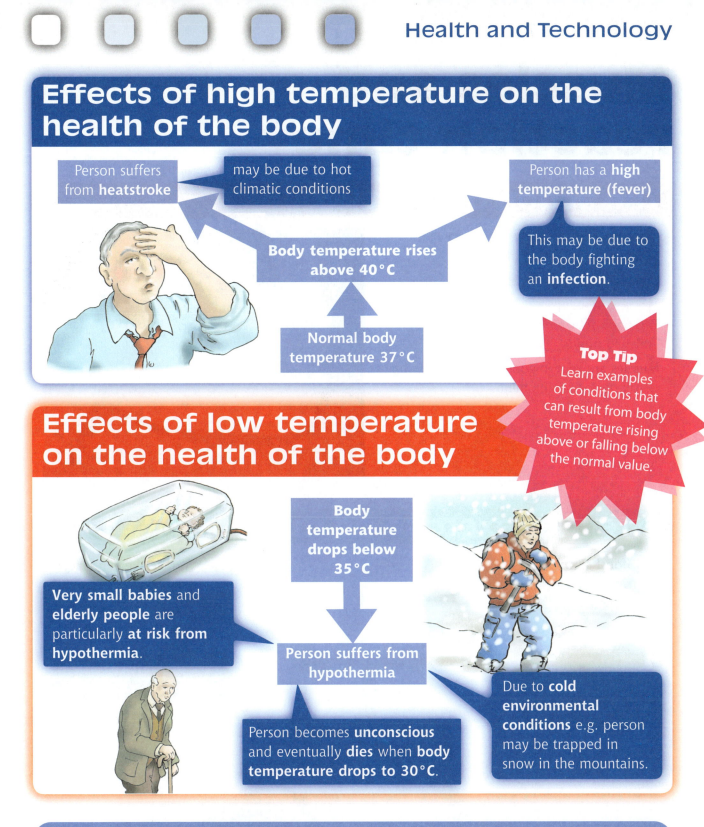

Person suffers from **heatstroke**

may be due to hot climatic conditions

Person has a **high temperature (fever)**

This may be due to the body fighting an **infection**.

Body temperature rises above 40°C

Normal body temperature 37°C

Top Tip
Learn examples of conditions that can result from body temperature rising above or falling below the normal value.

Effects of low temperature on the health of the body

Body temperature drops below 35°C

Very small babies and **elderly people** are particularly **at risk from hypothermia**.

Person suffers from hypothermia

Due to **cold environmental conditions** e.g. person may be trapped in snow in the mountains.

Person becomes **unconscious** and eventually **dies** when **body temperature drops to 30°C**.

Quick Test

1. Name **three** types of thermometer used to measure body temperature.
2. State the normal body temperature.
3. What can cause the temperature of the body to rise above 40°C?
4. What is hypothermia?
5. Name **two** groups of people that are particularly at risk from hypothermia.

Answers 1. Mercury glass clinical, digital clinical, liquid crystal. **2.** 37°C. **3.** Very warm climatic conditions (causes heatstroke), body may be fighting a viral infection. **4.** Condition of the body when its temperature drops below 35°C. **5.** Very young babies, elderly people.

Muscles and reaction time

The action of muscles

Movement of our body is only possible because of the **action of muscles**.

Our muscle system is **attached to a bony skeleton**.

The **contraction of muscles** enables our **limbs to move**.

Muscles are kept in good working order by **regularly exercising** them.

Effect of exercise on muscles

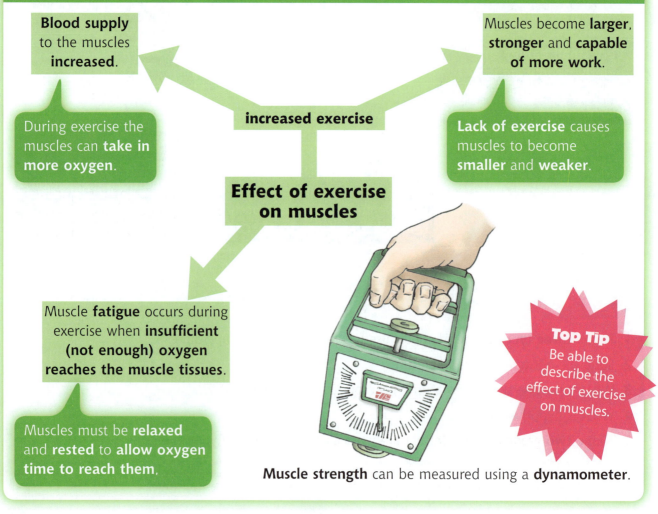

Blood supply to the muscles **increased**.

Muscles become **larger, stronger** and **capable of more work**.

During exercise the muscles can **take in more oxygen**.

increased exercise

Lack of exercise causes muscles to become **smaller** and **weaker**.

Effect of exercise on muscles

Muscle **fatigue** occurs during exercise when **insufficient (not enough) oxygen reaches the muscle tissues**.

Muscles must be **relaxed** and **rested** to **allow oxygen time to reach them**.

Top Tip
Be able to describe the effect of exercise on muscles.

Muscle strength can be measured using a **dynamometer**.

Reaction time

Reaction time is the **time taken to respond (react) to a stimulus**.

Reaction time can be **measured** using

● **a dropped ruler**   ● **an electronic timer**.

The time to react when a **light is switched on** or a **sound is made** is measured using an **electronic timer**.

The **length of fall of a ruler** is measured.

Reaction times are worked out using a chart.

Investigation: Investigate the **effect of practice on reaction time**.

Factors that affect reaction time

A person will take longer to react when under the influence of drugs or alcohol.

drugs ← **Factors that affect reaction time** → **alcohol**

practice   **excitement**

Top Tip
Remember the main factors that affect reaction times. Be able to give an example of a condition that a person may be suffering from if their reaction time was long.

Reaction time as an indicator of health

A **long reaction** time can indicate that a person may be suffering from:

● **brain or nerve disorders**   ● **arteriole disease**   ● **diabetes.**

Quick Test

1. Explain why muscles are important.

2. What effect does regular exercise have on the blood supply to the muscles?

3. Describe the effect of lack of exercise on the strength of muscles.

4. What causes muscle fatigue?

5. What is meant by the term **reaction time**?

6. Name **three** factors that affect reaction time.

7. State a condition that a person may be suffering from if their reaction time was long.

Answers: 1. It is the contraction of muscles that causes our limbs to move. **2.** Blood supply to the muscles increases, enables muscles to take in more oxygen. **3.** Muscles become smaller and weaker. **4.** Not enough oxygen reaches the muscles during exercise. **5.** The time taken to respond (react) to a stimulus. **6.** Alcohol, drugs, excitement practice. **7.** Brain or nerve disorders, arteriole disease, diabetes.

Effects of alcohol on the body

Effects of alcohol

Any alcohol taken into the body is **absorbed very quickly through the stomach wall** and into the **bloodstream**.

Its effect on the brain can lead to:

- a **longer reaction time**
- **poor muscle control**
- **poor judgement**.

Other **drugs** can have a **similar effect**.

All of these increase the risk of an accident.

Top Tip
Remember the short term effects of alcohol on the brain.

Measuring the alcohol content in exhaled air

This can be **measured** using

- a **breathalyser**
- an **alcometer**

This is used by the police if a person is suspected of **'being over the limit'**.

Health risks related to alcohol and drugs

Taking drugs during pregnancy can **damage the health** of the baby.

Drugs should **only be taken by a pregnant woman if prescribed by her doctor**.

Drinking excessive amounts of alcohol can cause **liver** and **brain** damage.

Long-term effect of excessive amounts of alcohol.

Top Tip
Be able to give examples of health risks that are related to the abuse of alcohol and drugs.

Quick Test

1. Describe **three** short-term effects of alcohol on the body.

2. Name **two** pieces of equipment that are used to measure the alcohol content of exhaled air.

3. Give an example of a **long-term** effect of drinking alcohol.

Answers 1. Longer reaction times, poor muscle control, poor judgement. 2. Breathalyser, alcometer. 3. Liver damage, brain damage.

Dairy industries – Milk

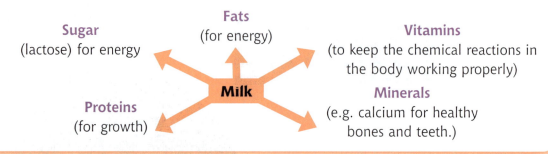

Nutritional value of milk

Milk is an important food in the diet of most people.

Sugar
(lactose) for energy

Fats
(for energy)

Vitamins
(to keep the chemical reactions in the body working properly)

Proteins
(for growth)

Milk

Minerals
(e.g. calcium for healthy bones and teeth.)

Different types of milk

Different **processing treatments** are used to produce **different types of milk**.

The **taste** of the milk is changed by the way it is treated.

Pasteurised

UHT
(Ultra High Treatment)

Evaporated

Semi-skimmed

Skimmed

Type of milk	Type of treatment
Pasteurised	Milk is heated to 72°C for 15 seconds and then quickly cooled This kills **most** of the harmful bacteria.
UHT	Milk is heated to a **higher** temperature (135–142°C) for 2–5 seconds. This kills **all** bacteria in milk preserving it and giving it a much longer shelf life. This treatment **changes the taste** of the milk.
Evaporated	Milk is heated to **remove some liquid**, making it **more concentrated** and **creamy**.
Semi-skimmed	Milk is treated to **remove some of its fat content**.
Skimmed	Milk is treated to **remove nearly all of its fat content**.

What causes milk to go sour?

Bacteria in milk feed on the milk **sugar** (lactose) and change it into **lactic acid**.

lactose
(a sugar in milk)

→ **bacteria** →

lactic acid
(causes milk proteins to clot and the milk goes sour)

Top Tip
Learn the treatments that are used to produce the different types of milk as these are sometimes asked for in exam questions.

Testing the quality of milk LO2 Activity

Microbial tests are carried out on milk to find out whether it is fit to drink or not.

The **resazurin test** is used to demonstrate the presence of bacteria in the milk.

Milk samples

To pass this learning outcome you must successfully carry out the resazurin test to test the quality of several milk samples.

Resazurin dye added

Colour examined after 20 minutes

The presence of **bacteria** causes the **resazurin** dye to **change colour** from

blue → pink → clear

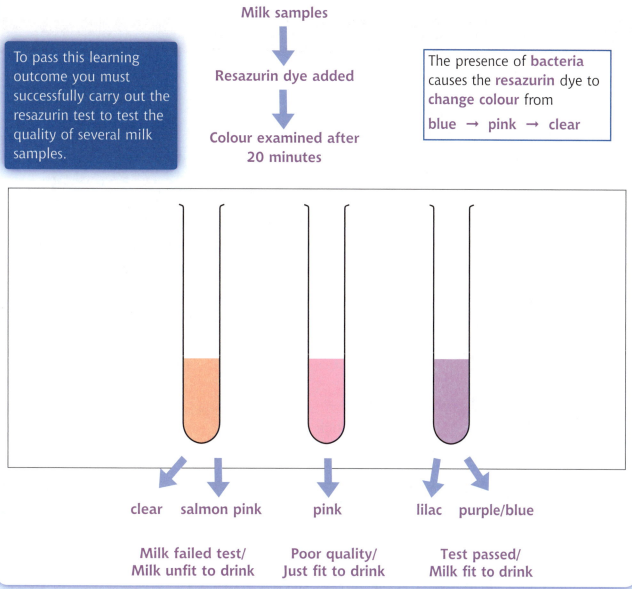

clear   salmon pink   pink   lilac   purple/blue

Milk failed test/   Poor quality/   Test passed/
Milk unfit to drink   Just fit to drink   Milk fit to drink

Quick Test

1. Name **three** components of milk that make it a good food.

2. Which type of milk has had nearly all of its fat removed?

3. Which type of milk has been heat treated to between 135°C and 142°C for 2–5 seconds to kill all bacteria?

4. Name the process by which milk is heated to 72°C for 15 seconds to kill harmful bacteria.

5. Name the test that is carried out on milk to check if it is fit for human consumption.

6. Describe the result of the test if the milk is unsafe to drink.

Answers 1. Protein, sugar, fat, vitamins, minerals. **2.** Skimmed **3.** UHT (Ultra High Treatment) **4.** Pasteurisation **5.** Resazurin test **6.** The dye becomes salmon pink or colourless.

Dairy industries – Yoghurt

Yoghurt bacteria

Yoghurt bacteria are examples of microbes (very small organisms that can be seen using a powerful microscope). They are added to pasturised milk to make yoghurt.

Milk **sugar** is converted into an **acid**. This thickens the milk and gives yoghurt its **flavour**.

Top Tip
Remember that the microbes used in the manufacture of yoghurt are bacteria.

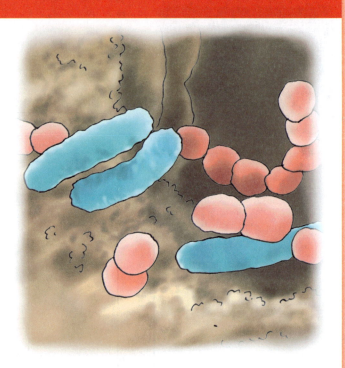

milk sugar →^{yoghurt bacteria} lactic acid

milk sugar —**yoghurt bacteria**→ lactic acid

Making yoghurt

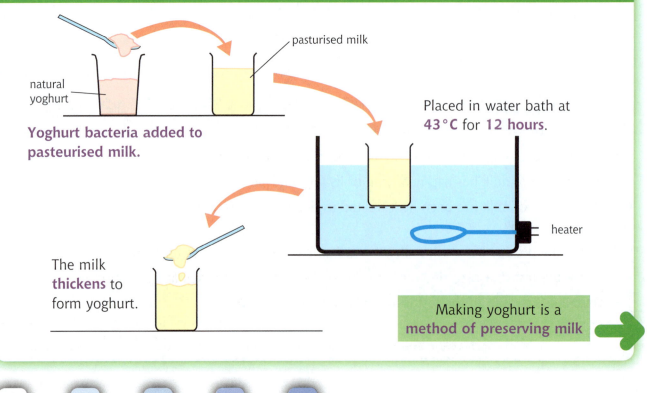

natural yoghurt

Yoghurt bacteria added to pasteurised milk.

pasturised milk

Placed in water bath at **43°C** for **12 hours**.

heater

The milk **thickens** to form yoghurt.

Making yoghurt is a **method of preserving milk**

Summary of the stages in the manufacture of yoghurt

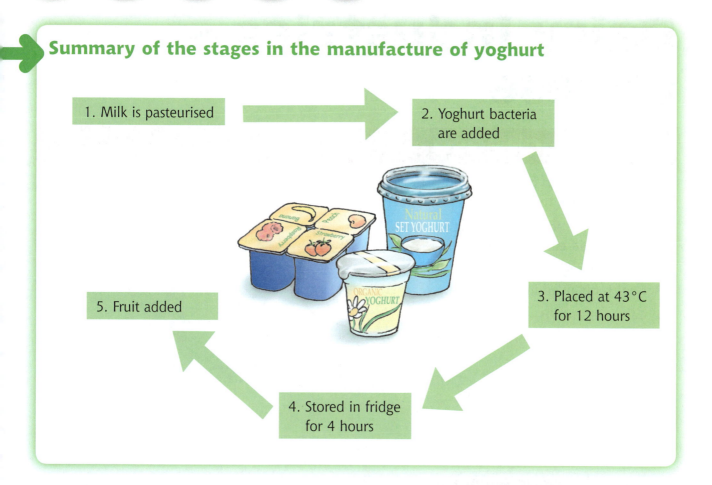

1. Milk is pasteurised

2. Yoghurt bacteria are added

3. Placed at 43°C for 12 hours

4. Stored in fridge for 4 hours

5. Fruit added

Quick Test

1. Which type of living organisms are used to make yoghurt?

2. What is milk sugar converted into during yoghurt production?

3. What effect does this substance have on the yoghurt?

Answers 1. Bacteria **2.** Lactic Acid **3.** It thickens the yoghurt and gives it flavour.

Dairy industries – Cheese

Making cheese

Rennet is added to **pasteurised milk** to cause the protein to clot.

The solid lumps formed when the protein clots are called **curds** and the liquid is referred to as **whey**.

Rennet is obtained from

- **Calves stomachs**
- **Genetically engineered fungi** grown in fermenters.

Pasteurised milk → Rennet → Curds + Whey

Salt is added to the chopped up curds.

The curds are **separated** from the whey and **pressed into a mould**.

The cheese is left for several months to **mature** and develop **flavour**.

Environmental impact of disposal of whey in rivers

Whey is a **waste product** of the cheese making industry.

If whey was disposed of directly into a river it would affect the **numbers of bacteria** and other **organisms such as fish**.

Location where water samples were tested	Oxygen concentration (units)
Disposal point	5
site 1	15
site 2	45
site 3	70
site 4	85

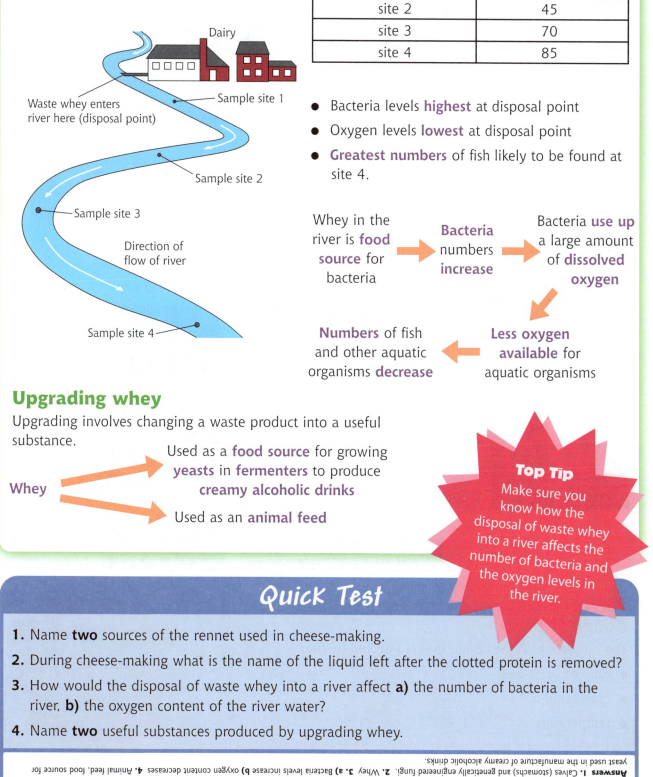

Dairy

Waste whey enters river here (disposal point)

Sample site 1

Sample site 2

Sample site 3

Direction of flow of river

Sample site 4

- Bacteria levels **highest** at disposal point
- Oxygen levels **lowest** at disposal point
- **Greatest numbers** of fish likely to be found at site 4.

Whey in the river is **food source** for bacteria → **Bacteria** numbers **increase** → Bacteria **use up** a large amount of **dissolved oxygen** → **Less oxygen available** for aquatic organisms → **Numbers** of fish and other aquatic organisms **decrease**

Upgrading whey

Upgrading involves changing a waste product into a useful substance.

Whey → Used as a **food source** for growing **yeasts** in **fermenters** to produce **creamy alcoholic drinks**

Whey → Used as an **animal feed**

Top Tip
Make sure you know how the disposal of waste whey into a river affects the number of bacteria and the oxygen levels in the river.

Quick Test

1. Name **two** sources of the rennet used in cheese-making.

2. During cheese-making what is the name of the liquid left after the clotted protein is removed?

3. How would the disposal of waste whey into a river affect **a)** the number of bacteria in the river, **b)** the oxygen content of the river water?

4. Name **two** useful substances produced by upgrading whey.

Answers 1. Calves (stomachs) and genetically engineered fungi. 2. Whey 3. a) Bacteria levels increase **b)** oxygen content decreases **4.** Animal feed, food source for yeast used in the manufacture of creamy alcoholic drinks.

Yeast-based industries – Bread

Making the dough

- Yeast is a **simple fungus** used in bread making.
- Dried yeast is mixed with flour, sugar and water to make a **dough**.
- If the bread dough is left in a warm place the yeast produces **carbon dioxide** gas which **causes the dough to rise**.
- The dough is then baked in an oven at a high temperature.

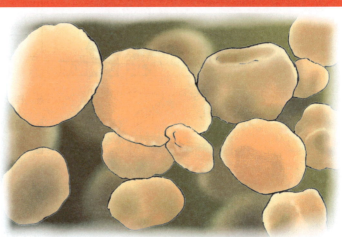

Factors that affect the rate at which the dough rises

The rate at which the dough rises is affected by the

- type of yeast used
- type of flour
- temperature.

Top Tip
Make sure you can name the gas produced by yeast which makes the bread dough rise.

Setting up experiments to investigate the factors that affect the rising of the dough

Carry out investigations using:

Investigations

- Different **types** of yeast

  Which **type of yeast** causes the dough to rise most after one hour?

- Different **temperatures**

  At which **temperature** does the dough rise most after one hour?

- Different **types** of flour

  Use **white** flour **brown** flour **wholemeal** flour

  Which dough rises the most after one hour?

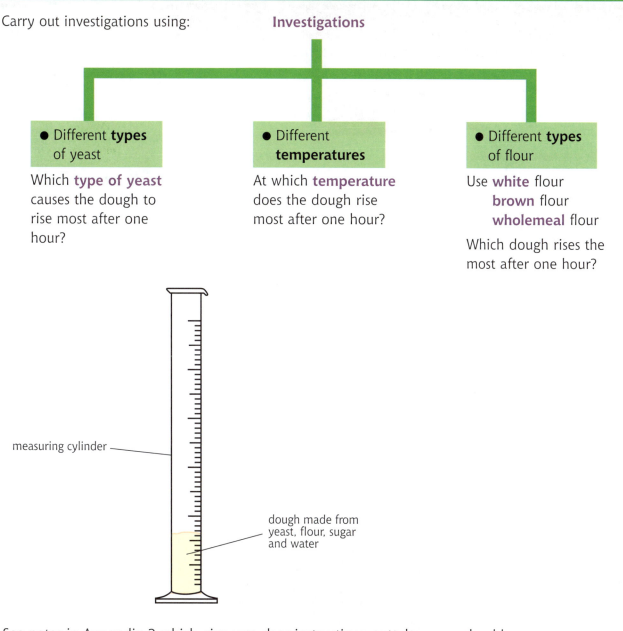

measuring cylinder

dough made from yeast, flour, sugar and water

See notes in Appendix 2 which give you clear instructions as to how you should write up an investigation report.

Quick Test

1. Name the type of micro-organism used in bread making.

2. Name the gas produced by yeast which causes the dough to rise.

3. Name **two** factors that affect the rising of the dough.

Answers 1. Yeast **2.** Carbon dioxide **3.** Temperature, type of yeast, type of flour.

Yeast-based industries – Beer

Making beer

The beer making industry makes use of the fact that yeast converts **sugar** into **alcohol** and **carbon dioxide**.

A single-celled fungus

yeast

sugar ⟶ alcohol + carbon dioxide

This process is called **fermentation**.

Factors that affect the alcohol content of the beer produced

The **temperature** at which the fermentation takes place

Factors that affect the alcohol content of the beer produced

The **type of yeast** used to ferment the sugars

The length of **time** that fermentation is allowed to take place

Investigations;
You could carry out investigations using **different types of yeast, different temperatures** and **different fermentation times**.

Cask-conditioned beer (real ale)

During the production of cask-conditioned beer the **yeast is not removed**. As a result, **fermentation** and **carbon dioxide production continue** in the cask.

The beer is **dark in colour** and **highly flavoured**.

The beer **does not last as long** as brewery beer, i.e. it **has a shorter shelf life**.

Top Tip
Be able to state the differences between cask-conditioned and brewery-conditioned beers.

Brewery-conditioned beer

During the production of brewery-conditioned beer, the **yeast is removed**. The fermentation **does not continue. Extra carbon dioxide is pumped into the beer** under pressure.

The beer is **clear, bright** like lager. The beer **lasts longer** than cask beer.

Quick Test

1. Write a word equation for the process of fermentation.
2. During the manufacturing process, name **three** factors that can affect the alcoholic content of the beer.
3. Why does fermentation continue in cask-conditioned beer?
4. State one difference in the production of brewery-conditioned beer and cask-conditioned beer.
5. Cask-conditioned beer has a long shelf life. True or false?

Answers 1. Sugar → alcohol + carbon dioxide. **2.** Temperature, type of yeast, fermentation time. **3.** Yeast is still present / has not been removed. **4.** Yeast is removed which stops the fermentation process in brewery beer. **5.** False.

Yeast-based industries – Fermented milk drinks

Making fermented milk drinks

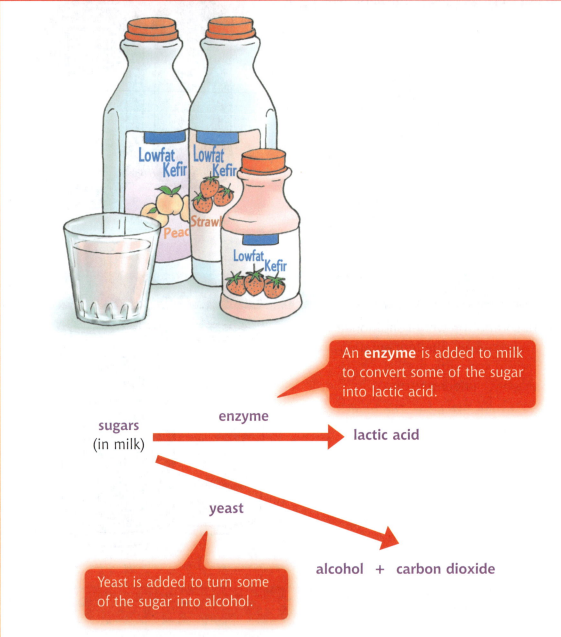

An **enzyme** is added to milk to convert some of the sugar into lactic acid.

sugars
(in milk) —— enzyme ——▶ lactic acid

yeast

Yeast is added to turn some of the sugar into alcohol.

alcohol + carbon dioxide

An **alcoholic milk drink** called **Kefir** is made in this way

It is a **refreshing**, **slightly alcoholic yoghurt drink**.

Immobilisation of yeast in the production of fermented milk drinks LO2 Activity

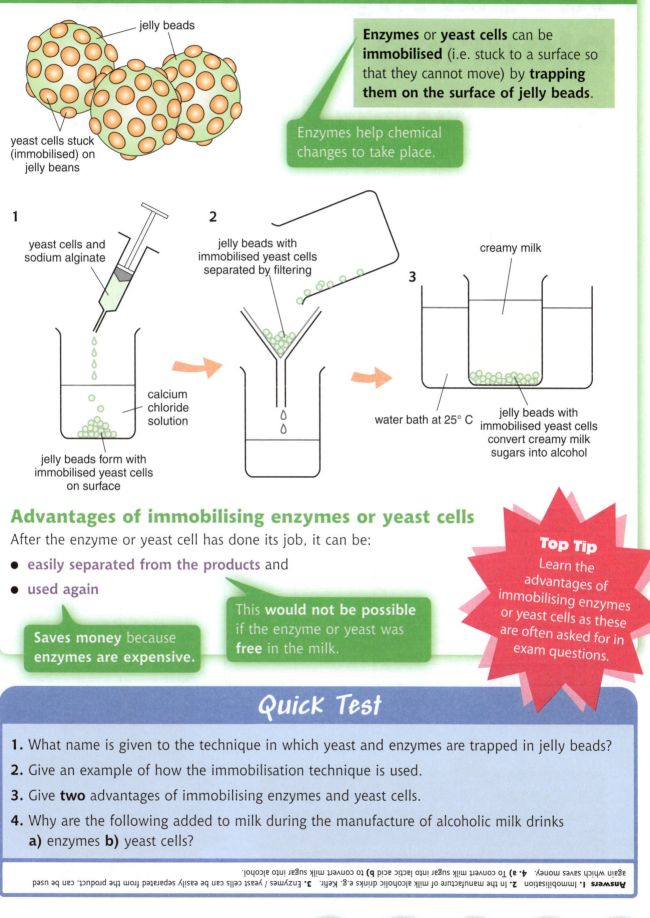

jelly beads

yeast cells stuck (immobilised) on jelly beans

Enzymes or yeast cells can be **immobilised** (i.e. stuck to a surface so that they cannot move) by **trapping them on the surface of jelly beads**.

Enzymes help chemical changes to take place.

1

yeast cells and sodium alginate

calcium chloride solution

jelly beads form with immobilised yeast cells on surface

2

jelly beads with immobilised yeast cells separated by filtering

3

creamy milk

water bath at 25° C

jelly beads with immobilised yeast cells convert creamy milk sugars into alcohol

Advantages of immobilising enzymes or yeast cells

After the enzyme or yeast cell has done its job, it can be:

- **easily separated from the products** and
- **used again**

This **would not be possible** if the enzyme or yeast was **free** in the milk.

Saves money because **enzymes are expensive**.

Top Tip
Learn the advantages of immobilising enzymes or yeast cells as these are often asked for in exam questions.

Quick Test

1. What name is given to the technique in which yeast and enzymes are trapped in jelly beads?
2. Give an example of how the immobilisation technique is used.
3. Give **two** advantages of immobilising enzymes and yeast cells.
4. Why are the following added to milk during the manufacture of alcoholic milk drinks
   a) enzymes **b)** yeast cells?

Answers 1. Immobilisation **2.** In the manufacture of milk alcoholic drinks e.g. Kefir. **3.** Enzymes / yeast cells can be easily separated from the product, can be used again which saves money. **4. a)** To convert milk sugar into lactic acid **b)** to convert milk sugar into alcohol.

Yeast-based industries – Flavouring and food colouring

Making food flavourings

Yeast is used in the manufacture of substances (**flavourings**) that **improve the taste and smell** (flavour) of food, e.g. crisps.

An **experiment** can be carried out using **fresh yeast cake** to illustrate the processes which are used to produce different flavours.

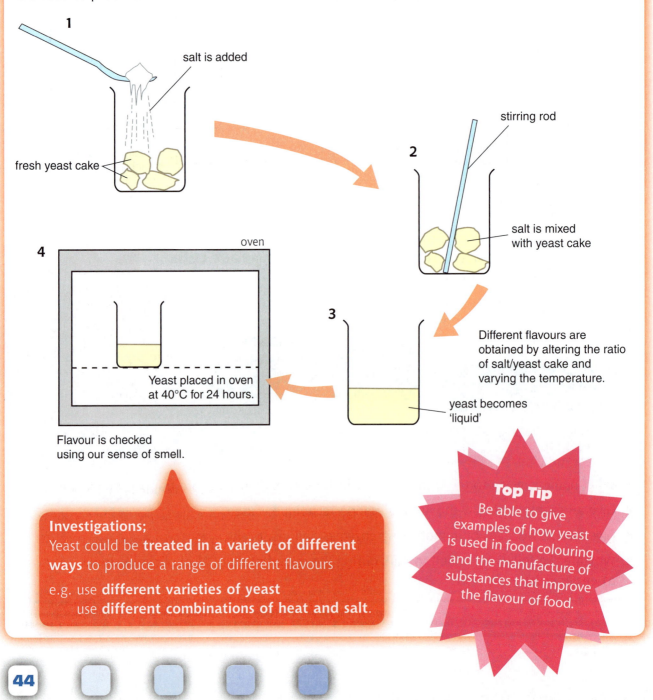

1
salt is added
fresh yeast cake

2
stirring rod
salt is mixed with yeast cake

3
Different flavours are obtained by altering the ratio of salt/yeast cake and varying the temperature.
yeast becomes 'liquid'

4 oven
Yeast placed in oven at 40°C for 24 hours.
Flavour is checked using our sense of smell.

Investigations;
Yeast could be **treated in a variety of different ways** to produce a range of different flavours

e.g. use **different varieties of yeast**
use **different combinations of heat and salt**.

Top Tip
Be able to give examples of how yeast is used in food colouring and the manufacture of substances that improve the flavour of food.

Food colouring

In the wild, salmon **feed on invertebrates,** e.g. **shrimps and prawns**

⬇

A **pink substance** in the shrimps and prawns causes the **flesh of the salmon to become pink**.

⬇

Salmon is **more attractive** to the consumer who **will want to buy and eat it**.

Artificially colouring salmon

Salmon produced in fish farms are fed on pellet food that **lacks the pink substance**.

£1.48

3 for £3.90
Lemon & Pepper
Salmon Fillet Portion

This problem is overcome by feeding the salmon with a **type of yeast which contains the same pink substance** that is found naturally in the shrimps and the prawns in the environment.

Salmon flesh is **grey in colour** and **unattractive to the consumer**.

Environmental impact of the waste from the yeast industries

Waste from yeast-based industries can have the **same effect** on rivers as whey, i.e. it would affect the **numbers of bacteria** and **other organisms such as fish**.

The waste can be

- upgraded to **cattle feed** (cattle cake)
- the yeast is used to **upgrade whey by converting sugars in the whey to alcohol** in the manufacture of creamy alcoholic drinks.

Quick Test

1. What name is given to a substance that improves the taste and smell of food?
2. Describe how fresh yeast cake is treated to produce different flavours.
3. What causes the flesh of salmon, reared in their natural environment, to look pink?
4. How does the flesh of salmon reared in fish farms differ from the flesh of salmon reared in the wild?
5. How is the flesh of salmon, reared in fish farms, made to look more attractive to the consumer?
6. Describe the environmental impact of waste from yeast-based industries on a river.
7. Give two ways in which waste from yeast-based industries is upgraded.

Answers 1. A flavouring. **2.** It is mixed with salt / stirred until it becomes a liquid / placed in an oven at 40°C for 24 hours. **3.** A pink substance which enters their body as a result of eating shrimps and prawns. **4.** The flesh of salmon in fish farms is grey / dull in colour. **5.** The salmon are fed on a type of yeast containing the same pink substance found in shrimps and prawns. **6.** Same effect as whey. It acts as a food source for bacteria > bacteria numbers increase > bacteria use dissolved oxygen > oxygen level decreases > numbers of fish and other aquatic life decrease. **7.** Waste is converted into cattle cake, used to upgrade whey during the manufacture of creamy alcoholic drinks.

Detergent industries – Detergents

What is a detergent?

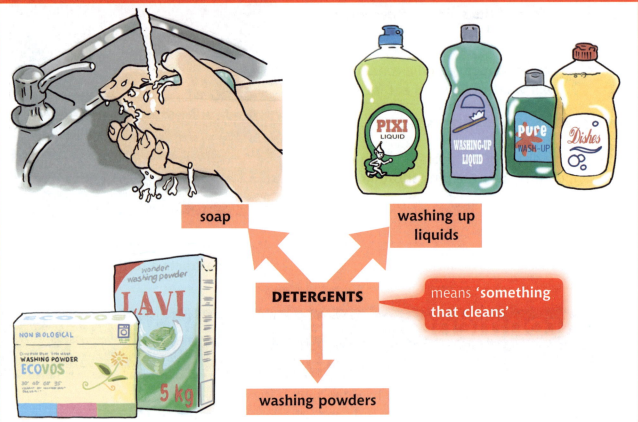

soap

washing up liquids

DETERGENTS — means 'something that cleans'

washing powders

The use of enzymes

Make up only **1%** of the powder

Enzymes digest (break down) stains at moderate temperatures

contains **enzymes**

This is overcome by enclosing the enzyme in a harmless coating

Biological washing powder

Are **produced by bacteria in fermenters**

Some people are **allergic** to biological powders

Top Tip
Be able to explain how enzymes work in biological detergents.

Non-biological washing powders **do not contain enzymes**

Stain removal using washing powders LO2 Activity

An experiment can be carried out to show the stain-removing ability of **biological** and **non-biological** washing powders at a **moderate temperature**

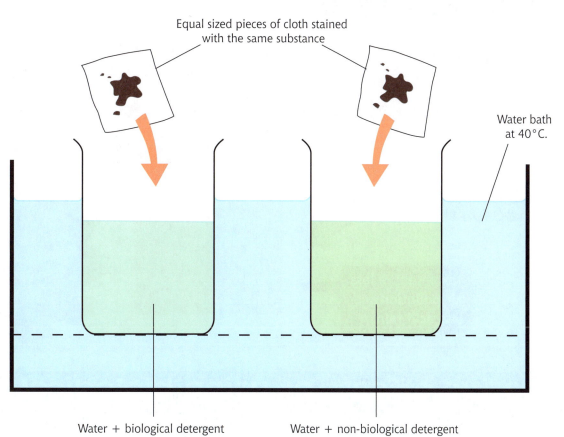

Equal sized pieces of cloth stained with the same substance

Water bath at 40°C.

Water + biological detergent

Water + non-biological detergent

The appearance of each piece of cotton tape is noted after 30 minutes.

Investigation

An investigation could be carried out to compare the stain-removing ability of a biological washing powder at

a) **moderate temperature** and b) **high temperature**.

At which temperature does the biological powder work best?

Quick Test

1. Name the substance found in biological washing powders that enables stains to be removed at moderate temperatures.

2. Name the organisms used to produce enzymes used in detergents.

3. The enzymes in biological washing powders are enclosed in a harmless coating. Suggest a reason for this.

Environmental impact of the detergent industries

Positive effects of using biological washing powders on the environment

Washing at lower **temperatures reduces the need for electricity to heat the water.**

Less fuel consumption

This, in turn, **reduces the amount of fossil fuel burned** at power stations to generate the electricity.

This results in **less air pollution** and **less acid rain**.

Top Tip
Learn about the positive and negative effects of biological detergents on the environment.

Negative effects of using biological washing powders on the environment

Detergents (in waste water) **contain chemicals** that can be **toxic** (**poisonous**) to wildlife.

Chemicals, e.g. **phosphates** in the powders can cause an **increase in the growth of algae** in lochs and rivers.

When the algae die, bacterial numbers **increase** as they feed on the dead remains. They **use up more dissolved oxygen in the water**. The **numbers of fish and other aquatic life decrease.**

This is called an **algal bloom**

Ways of reducing the effect of detergents on the environment

The effects of detergents on the environment can be reduced by:

- **decreasing the amount of chemicals** such as **phosphates** in the powders
- **removing these chemicals at water treatment plants** before releasing the water into the environment.

The advantages of using biological washing powders

The enzymes in the washing powder are **destroyed at temperatures above 60°C**.

Biological powders, therefore, **work best** at **lower temperatures (40° to 50°C)**.

This saves energy

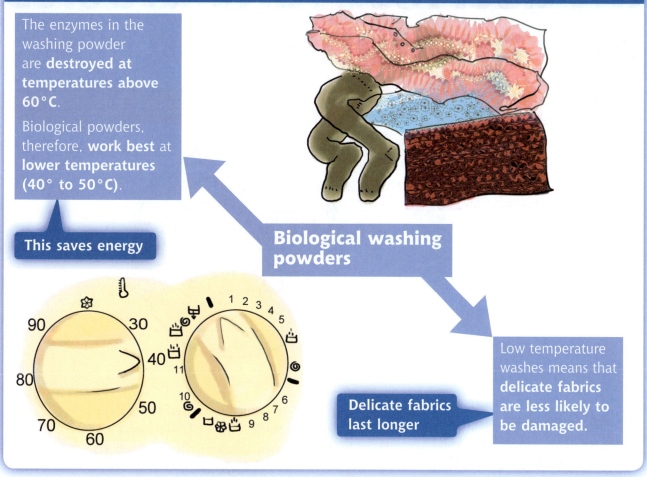

Biological washing powders

Low temperature washes means that **delicate fabrics are less likely to be damaged.**

Delicate fabrics last longer

Quick Test

1. Give **two** examples of damage to the environment which might be caused by detergents.

2. Name chemicals in detergent powders that cause an increase in the growth of algae in lochs and rivers.

3. What name is given to this increased growth of algae in aquatic environments?

4. Suggest **two** advantages of using biological detergents.

Answers 1. Some detergents are toxic to wildlife; can increase growth of algae > dead algae > increased number of bacteria > less dissolved oxygen > can reduce numbers of aquatic organisms. **2.** Phosphates **3.** An algal bloom **4.** Moderate temperature washes > saves energy and money and less damage caused to delicate fabrics.

Pharmaceutical industries – antibiotics and antifungals

Antibiotics

In 1928 **Alexander Fleming** discovered that a substance produced by a fungus **prevented the growth of bacteria**. He called the substance an **antibiotic**.

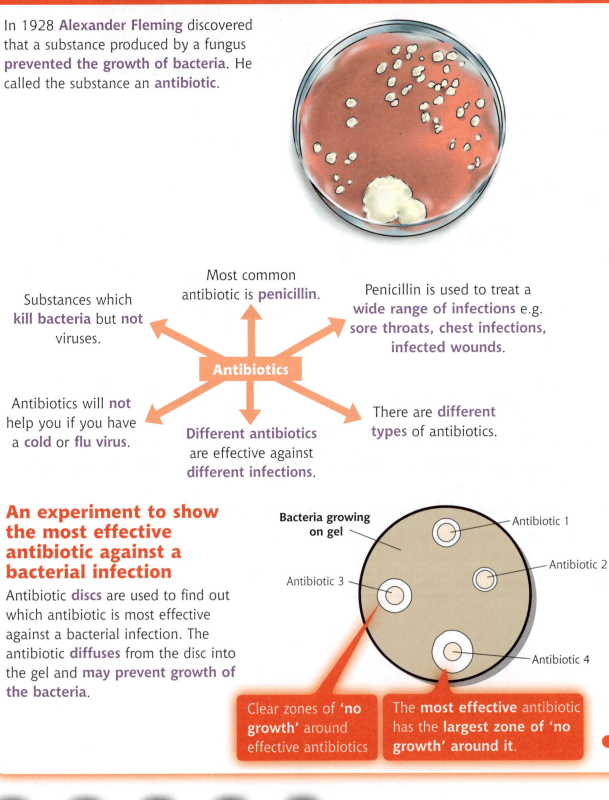

Substances which **kill bacteria** but **not** viruses.

Most common antibiotic is **penicillin**.

Penicillin is used to treat a **wide range of infections** e.g. **sore throats, chest infections, infected wounds**.

Antibiotics

Antibiotics will **not** help you if you have a **cold** or **flu virus**.

Different antibiotics are effective against **different infections**.

There are **different types** of antibiotics.

An experiment to show the most effective antibiotic against a bacterial infection

Antibiotic **discs** are used to find out which antibiotic is most effective against a bacterial infection. The antibiotic **diffuses** from the disc into the gel and **may prevent growth of the bacteria**.

Bacteria growing on gel

Antibiotic 1

Antibiotic 2

Antibiotic 3

Antibiotic 4

Clear zones of 'no growth' around effective antibiotics

The **most effective** antibiotic has the **largest zone of 'no growth' around it**.

Large scale production of antibiotics

Microbes are now **genetically engineered** to produce the desired antibiotics.

The **gene** that codes for the antibiotic is **transferred** to the microbe.

The microbe is then **grown rapidly in a fermenter** to produce **large quantities** of the antibiotic.

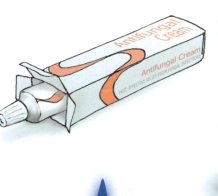

Top Tip
Be able to explain the difference between an antibiotic and an antifungal.

Antifungals

Some infections can be caused by **fungi**, e.g. **athlete's foot** and **thrush**.

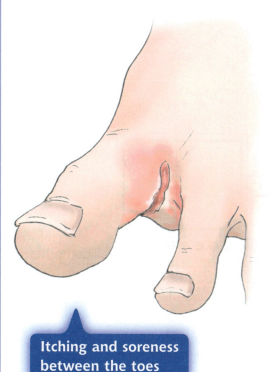

Itching and soreness between the toes

Antifungals are **chemicals** which **limit** or **stop** the growth of **fungal infections**

Thrush is a fungal infection which causes **white spots** to appear in the **mouth**.

It can be treated by using an **antifungal mouth wash**.

Quick Test

1. Name the scientist who first discovered penicillin.
2. Antibiotics are substances that kill viruses but not bacteria. Is this statement true or false?
3. Name a fungal infection of **a)** the feet **b)** the mouth.
4. What name is given to a substance which limits or stops a fungal infection?

Answers 1. Alexander Fleming **2.** False **3. a)** Athlete's foot **b)** thrush. **4.** An antifungal

Industrial production of antibiotics

The use of fermenters

It was around the time of the Second World War that scientists managed to produce penicillin on an **industrial scale**.

To meet the increasing demand for antibiotics, they are now produced in **large containers** called **fermenters**.

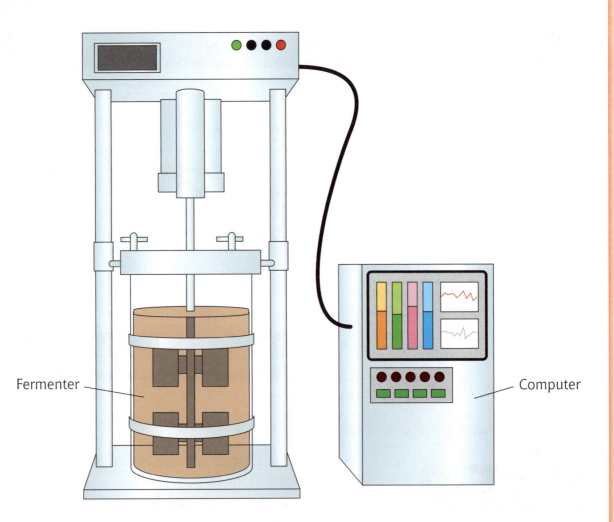

Fermenter

Computer

The conditions, e.g. **pH**, **temperature**, **oxygen level**, **nutrient levels**, in these fermenters are **carefully controlled (automated)** using **computers** so that the micro-organisms have the **best conditions** for multiplying and producing antibiotics.

Sterile conditions must be maintained inside the fermenter

The antibiotic has to be **purified** and **separated (extracted)** from other substances in the fermenter.

Top Tip
Be able to list the conditions that need to be controlled during the production of antibiotics in a fermenter.

Environmental impact of the pharmaceutical industry

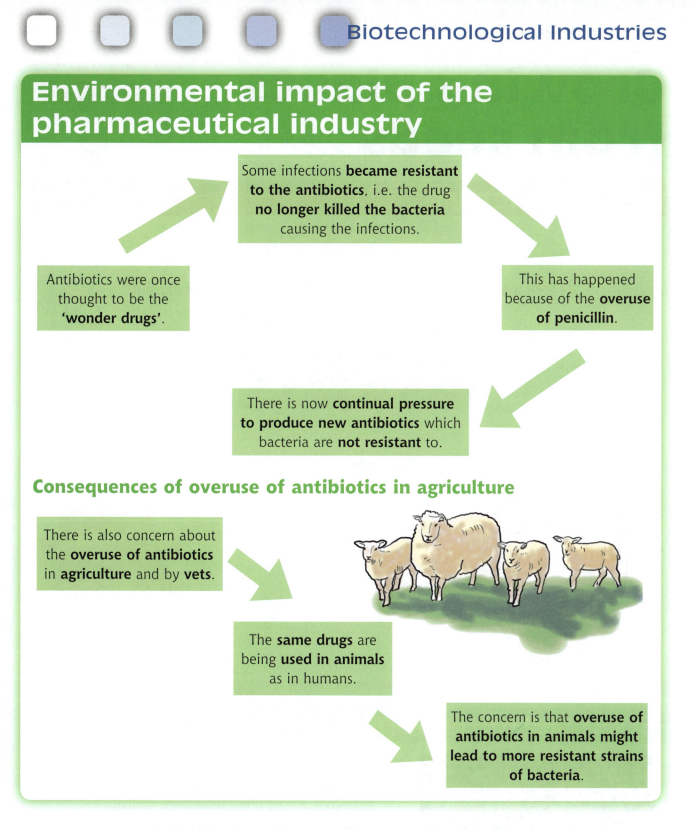

Some infections **became resistant to the antibiotics**, i.e. the drug **no longer killed the bacteria** causing the infections.

Antibiotics were once thought to be the **'wonder drugs'**.

This has happened because of the **overuse of penicillin**.

There is now **continual pressure to produce new antibiotics** which bacteria are **not resistant** to.

Consequences of overuse of antibiotics in agriculture

There is also concern about the **overuse of antibiotics** in **agriculture** and by **vets**.

The **same drugs** are being **used in animals** as in humans.

The concern is that **overuse of antibiotics in animals might lead to more resistant strains of bacteria**.

Quick Test

1. Name the large vessel used to produce antibiotics on an industrial scale.
2. State **two** conditions that need to be controlled in the vessel during the production of the antibiotics.
3. The overuse of antibiotics can lead to bacteria developing a resistance to the antibiotic. Explain what the term resistance means.

Growing plants from seeds

Seed structure

Seed coat

Embryo plant

Seed coat
(protects the seed)

Food store

Embryo root and
shoot (grows into
new plant)

Top Tip
Be able to
label a diagram
showing the main
structural parts
of a seed.

Seed germination

Germination is the development of the embryo root and shoot into a young plant.

During germination the **food reserves** in the seed are used up to provide **energy** for the growth of the root and the shoot.

Conditions for germination

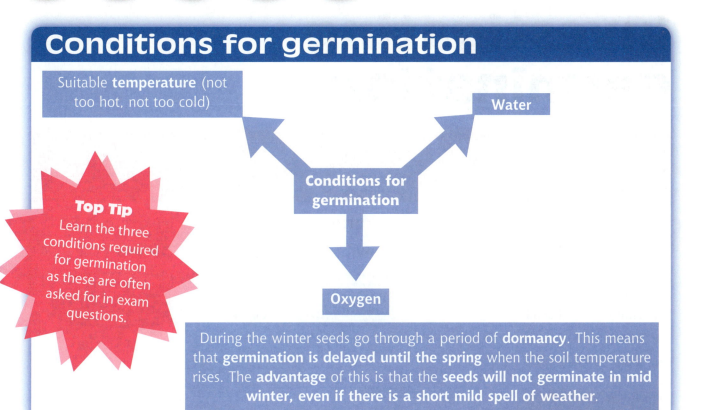

Suitable **temperature** (not too hot, not too cold)

Water

Conditions for germination

Oxygen

Top Tip
Learn the three conditions required for germination as these are often asked for in exam questions.

During the winter seeds go through a period of **dormancy**. This means that **germination is delayed until the spring** when the soil temperature rises. The **advantage** of this is that the **seeds will not germinate in mid winter, even if there is a short mild spell of weather**.

Photosynthesis

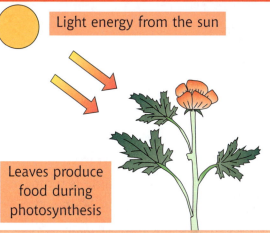

Light energy from the sun

Leaves produce food during photosynthesis

Photosynthesis is a process which takes place in **green plants**.

The plants use sunlight energy to **produce food in their leaves**.

The food produced by photosynthesis is **used by the plant for growth**.

Light energy is changed into **chemical energy**.

Quick Test

1. Name the part of the seed that gives protection.
2. Name the part of the seed that grows into a new plant.
3. Name the part of the seed that provides nutrients for the embryo to grow.
4. List **three** conditions required for germination.
5. Explain the advantage of seed dormancy.
6. After a seed has germinated the plant makes its own food using sunlight. Name this process of food production.

Answers 1. Seed coat **2.** Embryo plant. **3.** Food store. **4.** Water, warmth, oxygen. **5.** Seeds will not germinate in winter even if there is a short mild spell of weather. **6.** Photosynthesis.

Investigating germination

Investigating factors that may affect germination

To find out how varying the volume of water affects germination

The **only variable** is the **volume of water**.

- Same **growth medium**.
- Same **type** of seed.
- Same **spacing** of seeds.
- Same **temperature**.

0 ml water

5 ml water

10 ml water

15 ml water

20 ml water

How are you going to measure how much germination has taken place?

Remember: Repeating experiments makes the results **more reliable**. See Appendix 2 for more information on investigations

To find out how varying temperature affects germination

15 ml of water in each dish

20 cress seeds evenly spaced in each dish

Five identical dishes are set up and placed in five different temperature conditions, e.g. in freezer, fridge, room, propagator, etc.

same growth medium in each dish

The **only variable** is **temperature**.

- Same **growth medium**.
- Same **type** of seed.
- Same **spacing** of seeds.
- Same **volume** of moisture.

Top Tip
Make sure you know how to plan and set up an investigation as problem solving questions about investigations are asked in the exam.

Measuring how much germination has occurred

Seeds showing the **development of roots and shoots**.

This can be done by simply **counting the number of seeds that are developing roots and shoots.**

Root and shoot **lengths** are **measured** and **recorded**.

The extent to which germination has occurred is worked out by **measuring the length of the roots and/or shoots** and working out **average values**.

Quick Test

1. In an investigation to find the effect of temperature on germination **a)** name the one factor you would vary **b)** name **three** factors you would keep the same.

2. Describe **two** ways in which you could observe or measure how much germination has taken place.

3. Explain why repeating an experiment is important.

Answers 1. a) Temperature. **b)** Growth medium, type of seed, spacing of seeds, volume of water. **2.** Count the number of seeds per dish that showed any signs of shoot / root growth; Measure the length of shoots and or roots or a number of seeds from each dish and calculate average values. **3.** Repeating an experiment increases the reliability of the results.

Methods of sowing seeds

Instructions for sowing seeds

How seeds are sown **depends on their size**.

Seed packets give **instructions about the spacing of seeds** when planting them.

Sowing large seeds

Larger seeds are **easy to space out**.

Top Tip
Remember that the method of sowing seeds depends on their size. Be able to describe the different methods of sowing seeds.

Sowing small seeds

Smaller seeds can be **mixed with silver sand** before sowing. The silver sand 'dilutes' the seeds so that they can be spread more thinly.

Investigation: Compare the rate of germination in **pelleted** and **non-pelleted** seeds.

Pelleted seeds **take longer to germinate** and **need more water**.

Pelleted seeds are small seeds **enclosed in a ball of clay**.

Pre-germinating (chitting) seeds

Chitting involves treating seeds so that they start to **germinate before planting them**.

Seeds that have a **very tough outer coat fail to germinate** until water from the soil gets in. To **speed up** germination, the seeds are **cracked** or a **small cut** is made in the outer coat to allow water to enter.

Sometimes part of the hard coat is removed using a **fine abrasive sand paper**.

Pea seeds are pre-germinated by **soaking them** and **placing them on layers of damp paper**.

Quick Test

1. What is the best way to sow very small seeds?

2. What are pelleted seeds?

3. Give one advantage of sowing pelleted seeds.

4. How does the germination time vary between pelleted and non-pelleted seeds?

5. What do you call the process of germinating seeds before sowing?

Answers 1. Mix them with silver sand before sowing. **2.** Small seeds that are enclosed in a ball of clay. **3.** Small seeds can be more evenly spaced out. **4.** Pelleted seeds take longer to germinate than non-pelleted seeds. **5.** Pre-germination or chitting.

Sowing seeds

Stages in sowing seeds LO2 Activity

1. Select a **clean seed tray**, fill with **seed compost** and firm gently.

2. Mix seeds with **silver sand** if they are very small.

3. Scatter seeds thinly and evenly over the surface of compost.

4. Cover seeds with a **thin layer** of compost and **firm gently**.

5. **Water** the completed tray using a **watering can** with a fine rose.

To pass this learning outcome you must **successfully sow seeds and care for them until they germinate**.

6. Cover the seed tray with **cling film** and **place in a propagator**.

LO2 type questions

Questions on Learning Outcome 2 activities often involve **putting the stages of a process into the correct order**. A typical exam question is shown below.

The stages involved in sowing Petunia seeds are shown below.

1. Sprinkle compost over the seeds.

2. Spread the seeds evenly on the surface.

3. Fill the seed tray with compost and gently press to level it.

4. Water and cover the tray with cling film.

5. Press down the compost gently.

Which of the following shows the stages of sowing the seeds in the correct order?

Correct Answer

A   3 > 5 > 2 > 1 > 4

B   3 > 2 > 1 > 5 > 4

C   3 > 4 > 2 > 1 > 5

D   3 > 5 > 2 > 4 > 1.

See Appendix 1 pages 88–9 for other examples of LO2 type questions.

Vegetative Propagation

Natural propagation structures

Vegetative propagation is a method of producing plants which are **identical to the parent plant** and which makes use of structures such as **bulbs**, **tubers**, **runners**, etc. naturally formed by the plant.

Food storage organs – Bulbs

Bulbs have **side buds** that develop into **daughter bulbs**.

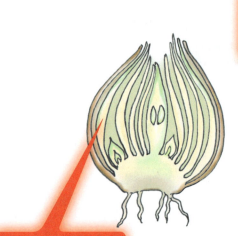

Swollen leaves containing **stored food**.

Food storage organs – Tubers

Swollen roots or **stems** containing **stored food**.

Root tubers can be **split** and grown as **separate** plants.

Natural plant propagation structures – Runners

Runner

Horizontal stem with plantlets at intervals along its length.

Natural plant propagation structures – Offsets

Small plantlet produced as a **side shoot at the base of the parent plant**.

Natural plant propagation structures – Plantlets

Mexican Hat Plant

Leaf plantlets

Leaf plantlets drop off and grow independently.

Quick Test

1. Name **two** food storage organs that are natural plant propagation structures.
2. What is a runner?
3. What name is given to a small plantlet that is produced as a side shoot at the base of the parent plant?
4. What method of propagation is used by the Mexican Hat plant?

Answers 1. Bulbs, root tubers. **2.** A runner is a horizontal stem with plantlets at intervals along its length. **3.** Offset. **4.** Leaf plantlets.

Artificial propagation – Cuttings

What is artificial propagation?

Artificial propagation methods are ways in which **plant growers produce new plants**, in a way in which plants themselves are unable to carry out.

A **node** is a point on the stem of a plant from which leaves or side branches grow.

Taking stem cuttings LO2 Activity

Top Tip
Be able to list the stages involved in taking cuttings in the correct order.

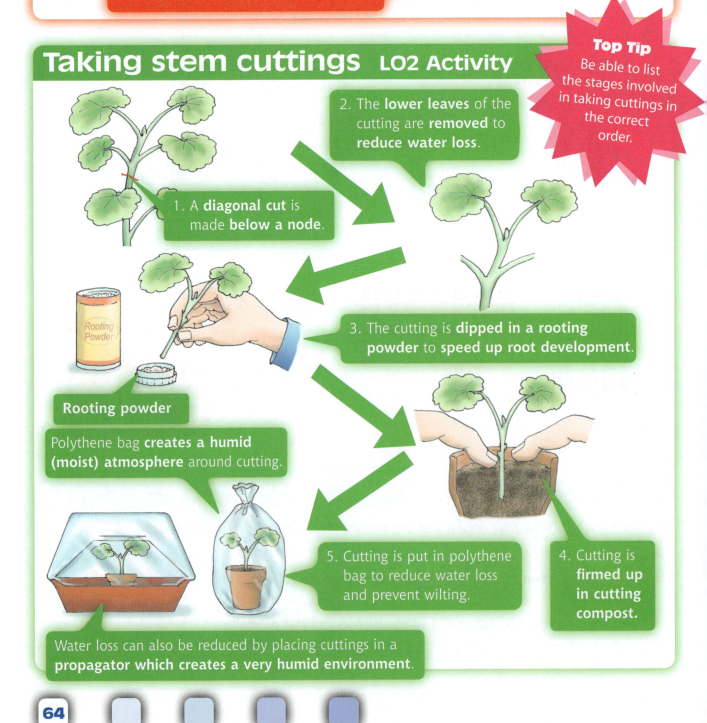

1. A **diagonal cut** is made **below a node**.

2. The **lower leaves** of the cutting are **removed** to **reduce water loss**.

3. The cutting is **dipped in a rooting powder** to **speed up root development**.

Rooting Powder

Rooting powder

Polythene bag **creates a humid (moist) atmosphere** around cutting.

4. Cutting is **firmed up in cutting compost**.

5. Cutting is put in polythene bag to reduce water loss and prevent wilting.

Water loss can also be reduced by placing cuttings in a **propagator which creates a very humid environment**.

Taking leaf cuttings

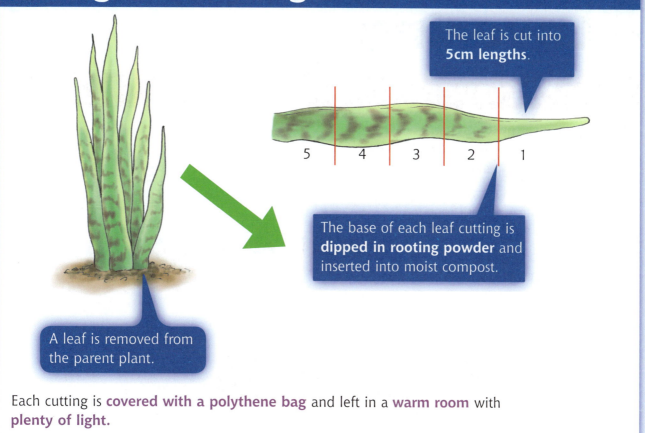

The leaf is cut into **5cm lengths**.

5 4 3 2 1

The base of each leaf cutting is **dipped in rooting powder** and inserted into moist compost.

A leaf is removed from the parent plant.

Each cutting is **covered with a polythene bag** and left in a **warm room** with **plenty of light**.

Quick Test

1. What is artificial propagation?
2. Give an example of an artificial propagation method.
3. What is a node?
4. Explain why the bottom leaves are removed when taking cuttings.
5. What is used to speed up the development of roots on a cutting?
6. Give **two** ways in which water loss and wilting are reduced when taking cuttings.

Answers 1. Ways in which plant growers increase the number of plants using methods plants cannot use. **2.** Taking leaf or stem cuttings. **3.** A point on a stem from which leaves or side branches grow. **4.** To reduce the surface area through which water can be lost from the cutting. **5.** Rooting powder. **6.** By placing the cuttings in a polythene bag or a propagator.

Artificial propagation – Layering

What is layering?

Layering involves **encouraging a stem or shoot to form roots** while still attached to the parent plant.

The stem is **pegged down in contact with the soil** until roots develop.

Some plants do this **naturally** when their stems touch the ground.

Stages in the layering process

A shoot is selected from the parent plant and its **lower leaves are removed**.

A **cut** is made on the **underside of the stem at a node**. This area is then **dusted with rooting powder**.

The side shoot is **pegged down in contact with the soil**.

Once the new plant has established its own roots, it should be cut free from the parent plant and potted out.

Advantages of layering

New plants are supported by getting nutrients and water from the parent plant.

The **success rate is greater** with plants that are **difficult to propagate from cuttings**.

Top Tip
Be able to state the advantages of propagating plants by layering.

The advantages and disadvantages of heat during propagation

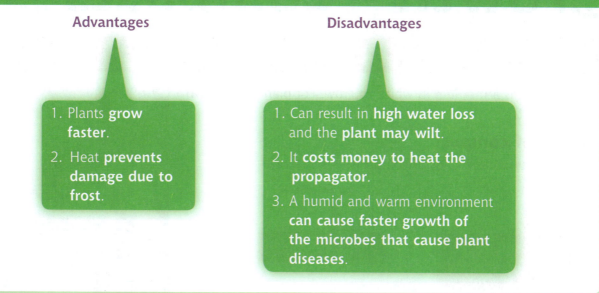

Advantages

1. Plants **grow faster**.

2. Heat **prevents damage due to frost**.

Disadvantages

1. Can result in **high water loss** and the **plant may wilt**.

2. It **costs money to heat the propagator**.

3. A humid and warm environment **can cause faster growth of the microbes that cause plant diseases**.

Quick Test

1. Plants can be propagated by layering. Explain what this process is.

2. List the stages involved in layering.

3. Give **two** advantages of layering.

4. Give **two** advantages and **two** disadvantages of using heat during the propagation of plants.

Answers: 1. This is the pegging down of stems into the soil until roots form at the nodes. **2.** A shoot is selected from the parent plant and its lower leaves are removed, a cut is made at a node and the area dusted with rooting powder, the side shoot is pegged down in contact with the soil. **3.** New plants are supported by getting nutrients and water from the parent plant, and the success rate is greater with plants that are difficult to propagate from cuttings. **4.** Advantages: plants grow faster, less frost damage. Disadvantages: high water loss / plants may wilt, higher energy costs, increased incidence of plant diseases.

Conditions for plant growth – Composts

Composts

A plant needs a **medium** in which to grow. Pot plants are grown in **composts**.

Types of compost

There are **two** basic kinds of composts:

Loam-based

Loamless

Contains **soil** as one of its components.

Contains **no** soil.

Each kind of compost contains a **mix of different substances to give the best conditions for plant growth**.

Composition of composts

	Loam-based compost	Loamless compost
Main component	● Loam (soil)	● Peat
Other substances added to improve growth	● Peat ● Sharp sand or perlite ● Fertilisers	● Sharp sand or perlite ● Fertilisers

Improves water-holding capacity of the compost.

The gritty sand improves aeration (volume of air getting into the soil) and **drainage** (allows excess water to pass between the soil particles more easily).

Increase the level of **plant nutrients** required for good growth.

Top Tip
Be able to list the components of compost and describe the importance of each.

Differences between a potting compost and a rooting compost

Some **fertiliser** is added to composts to give the plants the **nutrients** they require to grow healthily.

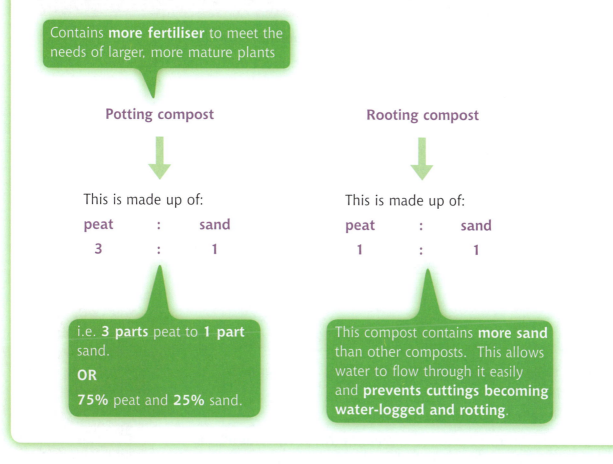

Contains **more fertiliser** to meet the needs of larger, more mature plants

Potting compost

This is made up of:

peat : sand
3 : 1

i.e. **3 parts** peat to **1 part** sand.

OR

75% peat and **25%** sand.

Rooting compost

This is made up of:

peat : sand
1 : 1

This compost contains **more sand** than other composts. This allows water to flow through it easily and **prevents cuttings becoming water-logged and rotting**.

Quick Test

1. Name the component of a compost that improves aeration and drainage.

2. Why is peat added to a compost?

3. Why is fertiliser important in a compost?

4. Name **two** types of compost.

5. a) Which type of compost has the higher percentage of sharp sand?

   b) Why is this important?

6. Why do potting composts contain more fertiliser than rooting composts?

7. Using the information about potting composts, calculate the volume of peat which has to be mixed with 10 litres of sand to make a potting compost.

Answers 1. Sharp sand or perlite. **2.** Improves water-holding capacity. **3.** Increases the level of plant nutrients required for good plant growth. **4.** Potting and rooting composts, loam, loamless. **5. a)** Rooting compost **b)** Prevents cuttings becoming water-logged and rotting. **6.** To meet the needs of larger, more mature plants. **7.** 30 litres of peat. (Ratio is 3:1 i.e. 3 x 10 : 1 x 10).

Conditions for plant growth – Fertilisers

Importance of fertilisers

Plants need certain chemicals (**nutrients**) for healthy growth.

The **nutrients** are taken into the plant from the **growth medium through its root system**.

The **main nutrients** contain **nitrogen (N), phosphorus (P)** and **potassium (K)**. These are referred to as **major plant minerals**.

Plants also need other minerals in much **smaller amounts** (**trace elements**) e.g. iron.

Mineral	Importance
Nitrogen (N)	**Leaf** growth
Phosphorus (P)	**Root** growth
Potassium (K)	Growth of **flowers/fruit**

Fertiliser ratios

There needs to be the correct **balance of nutrients** in the growth medium for plants to grow well.

This balance is called a **ratio**. (see Appendix 3)

The **proportions (relative amounts)** of nitrogen (N), phosphorus (P) and potassium (K) are always shown on the bag **in the order N**, **P** and **K**.

Top Tip
Be able to explain the term fertiliser ratio and carry out calculations related to this.

The bag contains:
  8%   nitrogen
  8%   phosphorus
  16% potassium.

A 100kg bag will therefore contain:
  8kg   of nitrogen
  8kg of phosphorus
  16kg of potassium.

Fertiliser
8 : 8 : 16
100kg

fertiliser ratio
8 : 8 : 16
(N) nitrogen
(P) phosphorus
(K) potassium

Applying fertilisers

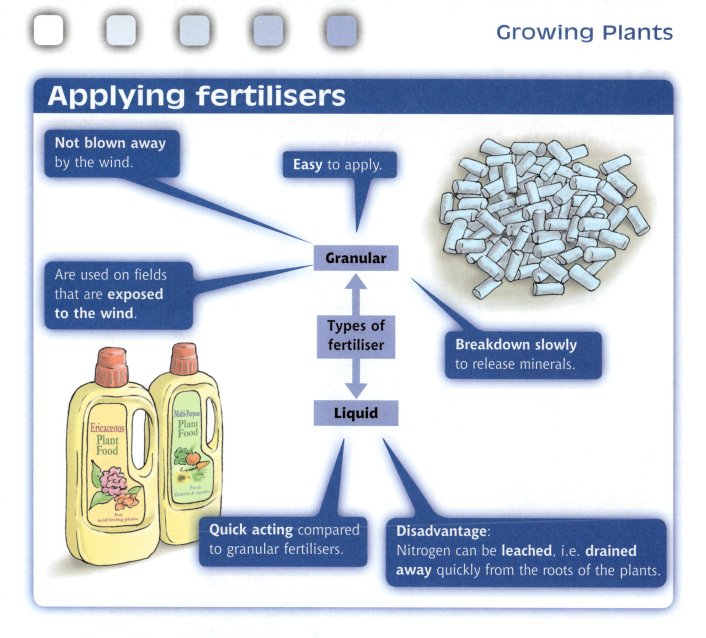

Not blown away by the wind.

Easy to apply.

Are used on fields that are **exposed to the wind**.

Granular

Types of fertiliser

Liquid

Breakdown slowly to release minerals.

Quick acting compared to granular fertilisers.

Disadvantage:
Nitrogen can be **leached**, i.e. **drained away** quickly from the roots of the plants.

Ericaceous Plant Food — For acid loving plants

Multi-Purpose Plant Food — For all flowers & vegetables

Quick Test

1. Name the **three** major plant minerals.
2. Which mineral promotes the development of **a)** roots **b)** fruit?
3. A fertiliser has a nutrient ratio of 7:7:14. To which plant nutrient does the 14 refer to?
4. How much potassium is in a 50kg bag of 8:5:12 fertiliser?
5. Give **two** advantages of using a granular fertiliser.
6. Give **one** disadvantage of using a liquid fertiliser.

Answers 1. Nitrogen, phosphorus, potassium. **2. a)** Phosphorus **b)** potassium. **3.** potassium **4.** 6kg **5.** Easy to apply, not blown away by the wind, breakdown slowly to release minerals. **6.** Can be leached away quickly from the roots of plants.

Watering plants 1

Importance of water to plants

Water is essential if plants are to grow and remain healthy.

Different plants require different amounts of water.

It is important that plants be given the **correct amount of water**.

Methods of watering plants

Trickle irrigation

Reservoir containing water.

Water from the reservoir reaches the individual pots through a plastic pipe.

Water **trickles** through pipes to reach each pot.

Capillary matting

One end of the matting is placed in a **reservoir** of water.

The matting **sucks up water** and **stays moist**.

Plants **absorb water from the moist matting**.

Capillary matting

Water reservoir

Top Tip
Be able to describe the different methods of watering plants.

Water retentive gels

♦ Releases water as plants need it
♦ Ideal for containers, bedding plants and vegetables
Works all seasons
Reduces watering by up to 75%

Gel granules

The gels are special chemicals that **absorb many times their own weight of water**.

When mixed with compost they are used in **planters** and **hanging baskets**.

This **prevents the containers drying out as quickly** and **reduces the need for watering**.

Quick Test

1. List **two** methods of watering plants while their owner is away from home on holiday.

2. What is a water retentive gel?

3. Give an example of how a water retentive gel is used to reduce the need to water plants.

Answers 1. Trickle irrigation, capillary matting. **2.** A chemical that is able to absorb many times its own weight of water. **3.** Mixed with compost in planters or hanging baskets.

Watering plants 2

Basic rules of watering LO2 Activity

Do not over-water.

Early watering is best

Basic rules of watering

To pass this learning outcome you must successfully water and look after a plant over a period of time. The plant should show no signs of over- or under-watering.

Some plants should be **watered from below**, e.g. African Violet

Do not water if plant is in direct sunlight.

Water droplets on a leaf can **concentrate the sun's heat rays** and **damage the leaf**

Top Tip
Learn the four basic rules of watering plants.

Signs of under-watering

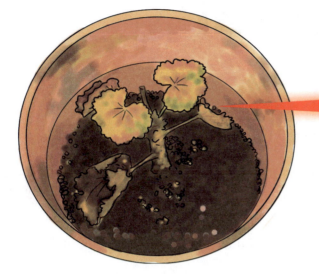

Leaf edges may become **brown** and **dry**.
Leaves may **fall off**.

Signs of over-watering

Compost may become **green**, **slimy** and **smelly**.
Leaves and stem may become **soft**, **yellow** and **decayed**.

Quick Test

1. List the **four** basic rules of watering plants.
2. Give **two** signs of **a)** over-watering **b)** under-watering plants.

Answers 1. Do not over-water, early watering is best, do not water if plant is in direct sunlight, some plants should be watered from below. **2. a)** Compost may become green, slimy and smelly, leaves may become soft, yellow and decayed. **b)** leaf edges may become brown and dry, leaves may fall off.

Controlling environmental conditions – Temperature

Optimum conditions

The environmental conditions in which a plant grows must be carefully controlled. Plants grow best within a range of temperatures referred to as the **optimum (best) conditions**.

Controlling environmental temperature

Maximum/minimum thermometers are used in greenhouses to record temperatures.

Maximum/minimum thermometer

A **thermostatically-controlled** heater is used to **control the temperature** in greenhouses and keep it within the **optimum range**.

Top Tip
Be able to describe how the temperature in a greenhouse is controlled and explain why this is important.

Maximum / Minimum thermometers

A max/min thermometer tells us **three** temperatures.

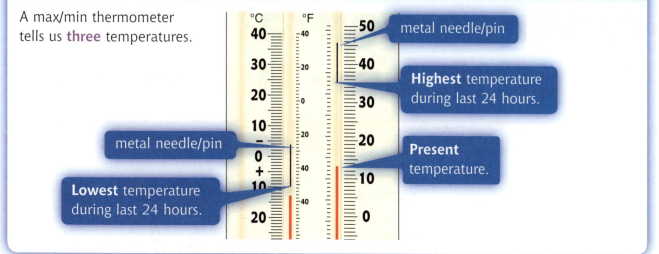

metal needle/pin

Highest temperature during last 24 hours.

metal needle/pin

Present temperature.

Lowest temperature during last 24 hours.

Protecting plants under glass

The plants are **protected** from **wind**, **rain** and **frost**.

Protecting crops in this way **raises the temperature of the soil**. Seeds can be **germinated earlier** in the season and **crops harvested earlier**.

Quick Test

1. Explain why it is important to control the environmental temperature in which a plant is growing.

2. What instrument is used to monitor the temperature in a greenhouse over a 24 hour period?

3. How is the temperature in a greenhouse kept as constant as possible?

Answers 1. Plants grow best within a range of temperatures referred to as the optimum range. **2.** Maximum/minimum thermometer. **3.** Using a thermostatically controlled heater.

Controlling environmental conditions – Humidity

What is humidity?

Humidity is the **amount of water vapour in the air**. This **varies from day to day** depending on the temperature and other climatic conditions. The humidity **inside** a greenhouse depends on the **temperature, ventilation, the number of plants** and **how much they are watered**.

A **wet and dry bulb thermometer** is used to measure humidity.

Ventilation

When plants are grown in greenhouses, **ventilation** is very important. Ventilation involves **providing a fresh air supply for the plants**.

A **fan** is used to **circulate air** in a greenhouse.

Effect of poor ventilation

Poor ventilation

↓

Stale air

↓

Ideal conditions for the spread of diseases

Diseased plant due to poor ventilation.

Top Tip
Be able to explain why good ventilation is important in a green house and list ways in which ventilation can be improved.

Automatic ventilation

Roof vents open and close automatically as the temperature of the greenhouse rises and falls throughout the day.

Louvred side panels can be used to aid ventilation. The louvres open and close automatically depending on temperature changes.

hot air out

cool air in

In large greenhouses fans are used to ensure good circulation of air.

Quick Test

1. What is meant by the term humidity?
2. Give **two** factors that affect humidity from day to day.
3. What piece of equipment is used to measure humidity levels?
4. Describe how poor ventilation in a greenhouse affects the health of plants.
5. Give **three** ways in which ventilation in a greenhouse can be improved.

Answers 1. The amount of water vapour present in the atmosphere. **2.** Temperature, rainfall. **3.** A wet and dry bulb thermometer. **4.** Poor ventilation > stale air > plants become diseased. **5.** Using a fan, automatic roof vents, automatic louvred side panels.

Plant maintenance

Needs of mature plants

Water

Light

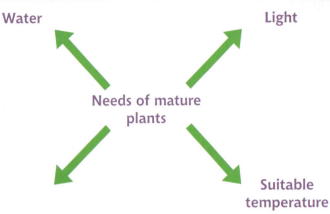

Needs of mature plants

Suitable temperature

N.B. The specific needs of mature plants vary, i.e. the needs of cacti are different from the needs of ferns, foliage plants or flowering plants.

Specific needs of cacti

Swollen stem

Leaves reduced to spines.

Warmth in summer/cool in winter.

Sunny/a lot of light.

Allow compost to dry between waterings.

Allow plenty of ventilation/do not mist.

Conditions cacti thrive in:

Specific needs of ferns

Average warmth/ferns do not grow well in hot conditions.

Bright light but not directly shining on them.

Compost must be moist at all times.

Mist regularly as ferns like humid conditions.

Specific needs of foliage plants

Average warmth.

Bright light.

Compost **moist** at all times.

Mist regularly.

> Plants grown for the **colour** and **shape** of their leaves.

Specific needs of flowering plants

Average warmth.

Bright light (some plants)/but not directly shining on them.

Compost **moist** at all times.

Mist regularly.

> Plants grown for their show of **colourful flowers**.

Top Tip

Learn about the specific needs of a variety of different types of plants and how these needs are related to their natural environment.

Quick Test

1. List the **four** needs of mature plants.

2. Describe the specific needs of cactus plants.

3. What are **a)** foliage plants **b)** flowering plants?

4. Describe **two differences** between the needs of cactus plants and ferns.

Methods of maintaining plants

Pricking out LO2 Activity

Pricking out is the **removal of seedlings** to **provide less crowded** growing conditions

1. **Select** a **clean tray** and **fill with seed compost**.

2. **Dig up a small clump of seedlings** and **separate them.**

3. Use a **dibber** to **make holes** in the compost and **plant seedlings separately**.

4. **Water** the completed tray using a **watering can with a fine rose.**

5. Tray of seedlings **after 2 weeks**.

To pass this learning outcome you must successfully complete stages 1 to 5.

Potting on LO2 Activity

This is the **transfer of a root bound plant into a larger pot**.

Roots **growing out of holes in the bottom of the pot**.

The plant is described as being **root bound** and needs to be replanted into a larger pot.

1. Carefully **remove plant from pot**.
2. **Select a pot** that is 25–30mm larger than the previous one.
3. **Place plant in the centre** of the new pot.
4. **Place potting compost around the edges** and **firm it down**.
5. **Water** the plant carefully.

To pass this Learning outcome you must successfully complete stages 1–5.

Dead heading

This is the **removal of dead flower heads**.

This **encourages the plant to continue to flower**.

Top Tip
Make sure you understand the importance of pricking out, potting on and dead heading as questions about these processes are asked in the exam.

Quick Test

1. Explain the term dead heading.
2. What effect does dead heading have on plant growth?
3. Why is pricking out important?
4. What would indicate that a plant needs potting on into a larger pot?

Answers 1. This is the removal of dead flower heads. **2.** It encourages the plant to continue to flower. **3.** Provides the seedlings with less crowded growing conditions. **4.** Roots growing out of holes in the bottom of the pot.

Methods of controlling pests and diseases

Plants under attack

Plants are **constantly under attack from organisms** that cause **disease** or **damage** them in some way.

Grey mould is a **common fungal disease** of plants. It occurs when plants are in **very damp environments** where the ventilation is poor.

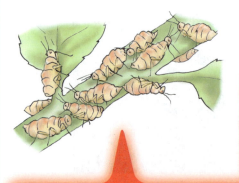

The **aphid** (greenfly) is an example of a **common plant pest**. Aphids damage plants by **feeding on the sugary sap** in the stems and leaves.

These **organisms are controlled** in a number of ways:

Top Tip
Be able to describe ways in which plants become damaged or diseased and describe how this is reduced.

Chemical control

Chemicals that **kill insects** that damage plants.

Insecticides

Pesticides

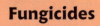

Chemicals that **kill invertebrate pests** that damage plants. Invertebrates are animals without back-bones, e.g. insects, slugs, snails.

Chemicals that **kill fungi** that damage plants.

Fungicides

Physical control

Burning affected plants

Burning **destroys diseased plants** and is sometimes necessary **to get rid of disease completely.**

Crushing pests

This **kills the pest** and **prevents any further damage** to plants.

Soapy water

This is used to **wash the glass in greenhouses** and **other equipment** that **might contain organisms that cause plant diseases**.

Biological control

Biological control occurs when one type of animal controls the numbers of another type of animal.

Ladybird (**predator**) **feeds on insect pest**, e.g. **aphid (greenfly)** and **prevents any further damage to plant**.

Quick Test

1. What is a fungicide?

2. Name **two** types of chemicals that are used to control invertebrate animals that damage plants.

3. Explain why the following methods of controlling plant pests and diseases are used,
   a) burning plants **b)** soapy water.

4. Give an example of the biological control of a plant pest.

5. Name a common plant disease and state how the incidence of the disease might be reduced.

Answers 1. A chemical that kills fungi that damage plants. **2.** Pesticides, insecticides. **3. a)** This destroys plants and is sometimes necessary to get rid of disease completely. **b)** this is used to wash the glass in greenhouses and other equipment that might contain organisms that cause plant diseases. **4.** The use of ladybirds to feed on the aphids (greenfly). **5.** Grey mould, reduced by making sure that plants are not too damp and that ventilation is good.

Protected cultivation

Protecting plants from weather and pests

Plants are **protected** from **wind**, **rain**, **frost** and **pests** in the following ways:

> Polythene fleeces allow the **water to pass through** and are light enough to rise with the plants as they grow.

Temperature is regulated by **thermostatically-controlled heater**.

Greenhouse

Polythene fleeces

Top Tip
Be able to describe ways in which plants can be protected and list the advantages of these.

How plants are protected from the weather and pests

Polythene tunnels

A **cloche** is a small polythene tent.

Advantages of using these methods to protect crops

Protecting crops in this way **raises the temperature of the soil**. Seeds can be **germinated earlier** in the season and **crops harvested earlier**.

A **larger crop yield** is also obtained due to **protection from the weather** and **pests**.

Quick Test

1. Describe **two** ways in which plants can be protected from very low temperatures.

2. List the advantages of protecting plant crops from cold conditions.

3. Apart from climatic conditions, what else are crops protected from by using any of the above methods?

4. How do these crop protection methods affect the crop yield?

Answers 1. By growing plants in greenhouses, polythene tunnels, cloches or under a polythene fleece. **2.** Seeds germinate earlier in the season and crops can be harvested earlier. **3.** Protected from pests e.g. insects. **4.** The crop yield is larger.

Appendix 1 – Types of Question

Knowledge and Understanding

You will be tested on your ability to recall what you have learned during the course and your understanding of facts and principles.

Some examples of multiple choice questions from past papers are shown below.

Top Tip
These are knowledge questions and can only be answered if you learn your course notes.

Example 1

Which of the following produces the carbon dioxide gas that makes dough rise?

A Bacteria
B Alcohol
C Yeast ◀ Correct answer
D Lactic acid

SQA 2004 Paper

Example 2

Immobilisation techniques can be used in the production of

A bread
B beer
C cheese
D fermented milk drinks ◀ Correct answer

SQA 2004 Paper

Example 3

The plant in the diagram needs to be

A dead headed ◀ Correct answer
B watered
C potted on
D given fertiliser

SQA 2004 Paper

Example 4

The list below contains some of the steps in the procedure for using a clinical thermometer.
1 Record the temperature accurately.
2 Remove the thermometer from the mouth.
3 Clean the thermometer with alcohol.
4 Place the thermometer under the tongue for two minutes.
5 Reset the mercury column.

Which of the following shows the correct sequence of steps?

A $5 \rightarrow 4 \rightarrow 1 \rightarrow 2 \rightarrow 5 \rightarrow 3$
B $1 \rightarrow 5 \rightarrow 4 \rightarrow 2 \rightarrow 1 \rightarrow 3$
C $3 \rightarrow 5 \rightarrow 4 \rightarrow 2 \rightarrow 1 \rightarrow 3$ ◀ Correct answer
D $4 \rightarrow 3 \rightarrow 2 \rightarrow 1 \rightarrow 5 \rightarrow 3$

SQA 2001 Paper

Top Tip
Examples 4 and 5 are testing two of the Learning Outcome 2 activities.

Example 5

A student is going to take a stem cutting from a geranium plant.

She uses secateurs to take a cutting which contains a growing point as shown in the diagram below.

The list below contains some of the remaining steps in the procedure.

1 Remove the lower leaves.

2 Place the cutting in the compost.

3 Water the compost.

4 Cut below a node.

5 Dip the base of the cutting into rooting powder.

Which of the following shows the correct sequence of steps?

A $4 \rightarrow 5 \rightarrow 2 \rightarrow 1 \rightarrow 3$

B $5 \rightarrow 4 \rightarrow 3 \rightarrow 2 \rightarrow 1$

C $5 \rightarrow 1 \rightarrow 4 \rightarrow 2 \rightarrow 3$

D $4 \rightarrow 1 \rightarrow 5 \rightarrow 2 \rightarrow 3$ ◁ Correct answer

SQA 2000 Paper

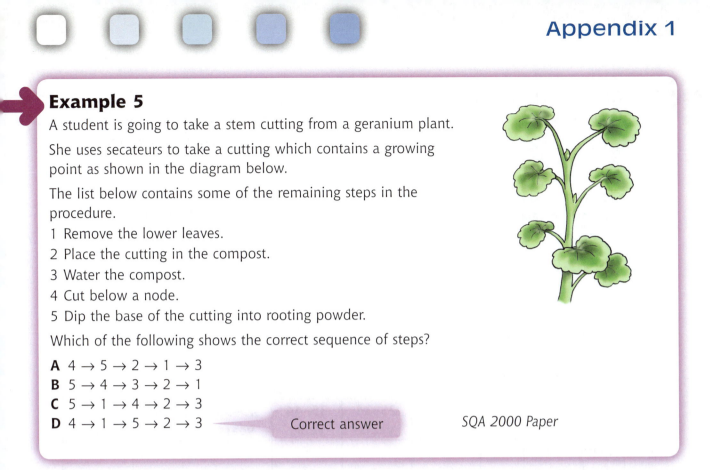

Problem Solving and Practical Abilities

Example 1

Seeds can be sown on capillary matting or in pots. These methods of sowing seeds were compared using three types of Cyclamen: *C. balearicum*, *C. coum* and *C. libanoticum*.

Ten seeds of each type were sown in pots and fifteen seeds of each type were sown on capillary matting.

C. balearicum took 89 days to germinate in pots and 120 days on capillary matting.

C. coum took 51 days to germinate in pots and 55 days on capillary matting.

C. libanoticum took 57 days to germinate in pots and 62 days on capillary matting.

(a) Use this information to complete the table by

   (i) providing headings

  (ii) putting in the results for each type.

This question asks you to **select relevant information from the text** and **put it in the table**.

Type of Cyclamen	Time to germinate (days)	
C. balearicum		
C. coum		
C. libanoticum		

(b) Which type of Cyclamen was the slowest to germinate?

> This question asks you to **draw a conclusion from the information in the table**.

_____

(c) Suggest one way in which the experimental procedure could be improved to make the comparison valid.

> This question asks you to **identify a way of improving an experimental procedure**.

_____

_____

SQA 2004 Paper

Example 2

(a) The table below shows the percentage of men in different age groups who are light, medium and heavy smokers.

	Percentage of men					
	16–24 years	25–34 years	35–44 years	45–54 years	55–64 years	65–74 years
Light smokers	13	7	6	3	2	4
Medium smokers	19	17	11	13	13	8
Heavy smokers	6	15	18	17	17	8

(i) What percentage of men aged 45–54 years are heavy smokers?

> This question asks you to **select information from the table**.

_____ %

(ii) Calculate the percentage of men aged 16–24 years who do not smoke.

> This question asks you to **analyse the data** and **calculate a value**.

_____ %

SQA 2004 Paper

Example 3

A student carried out an investigation to compare how well a washing powder worked at different temperatures.

The results are shown in the table below

Temperature (°C)	Percentage of stain removed
10	80
20	88
30	94
40	92
50	82
60	63

(a) On the grid below complete the **line graph** by

    (i)   providing a label for the horizontal axis

    (ii) completing the scale on the vertical axis

    (iii) plotting the remaining results.

> This question asks you to **present information in the form of a line graph.**

SQA 2004 Paper

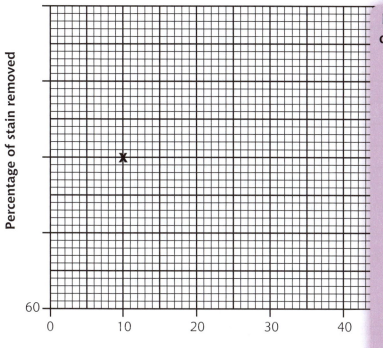

Remember the following when drawing graphs:

- Points that are plotted **must be joined up** to gain the mark.
- The line joining the plots **must go through the plot mark** or the mark for the plot will be lost.
- Make sure the **units** (where relevant) **are included** when adding a label to an axis.
- Do **not** extend the graph to zero if there are no values for zero in the data to be plotted.
- Make sure you choose a scale so that you use more than half of the graph paper.

Example 4

(a) Forty adults were asked to name their favourite drink.
The results are shown below.

SQA Paper 2004

Favourite drink	Number of adults
Beer	20
Lemonade	10
Wine	5
Water	5

Present the information in the table in the form of a pie chart.

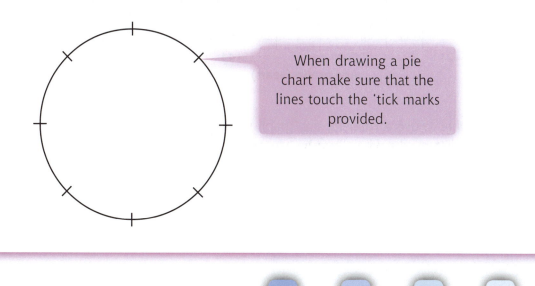

> When drawing a pie chart make sure that the lines touch the 'tick marks provided.

Example 5

In 1995 the recommended maximum drinking levels were 21 units of alcohol per week for men and 14 units for women.

The percentage of the population in each age group drinking more than the recommended maximum level is shown in the table below.

Age group (years)	Proportion of the population drinking more than the recommended maximum level (%)	
	Males	Females
16–24	37	18
25–34	34	15
35–44	33	12
45–54	31	12

(Scottish health survey 1995)

(a) On the grid below, complete the bar graph by

    (i) putting a scale on the vertical axis

    (ii) plotting the remaining results for males.

SQA Paper 2002

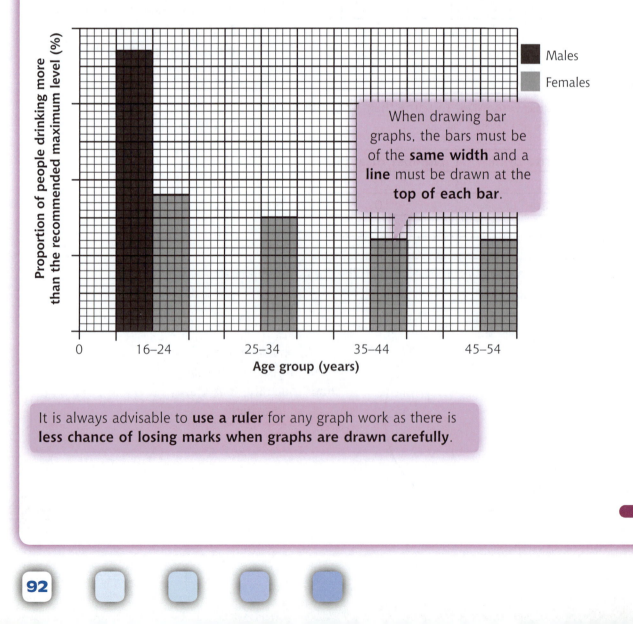

When drawing bar graphs, the bars must be of the **same width** and a **line** must be drawn at the **top of each bar**.

It is always advisable to **use a ruler** for any graph work as there is **less chance of losing marks when graphs are drawn carefully**.

Example 6

Question 1 and 2 refer to the investigation below.

The following dishes were set up to investigate different conditions affecting the germination of cress seeds.

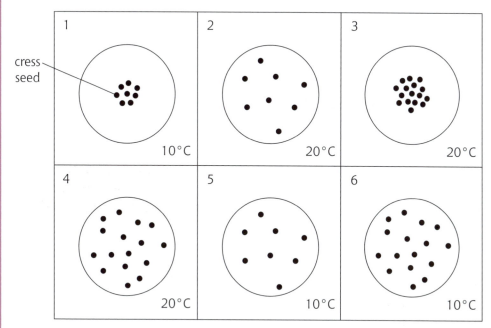

1. Which **two** dishes should be compared to find the effect of temperature on germination?

    A 1 and 2
    B 1 and 3
    C 3 and 5
    D 4 and 6

    Correct answer

2. To investigate the effect of spacing on germination, dish 3 is best compared with

    A dish 2
    B dish 4
    C dish 5
    D dish 6

    Correct answer

These questions ask you to **identify variables** i.e. the **one factor that should be altered** and the **other factors that should be kept constant**.

SQA 2004 Paper

Appendix 2 – Investigations (Learning Outcome 3)

This involves carrying out an investigation related to one of the units. The stages that must be carried out to pass this learning outcome are summarised below.

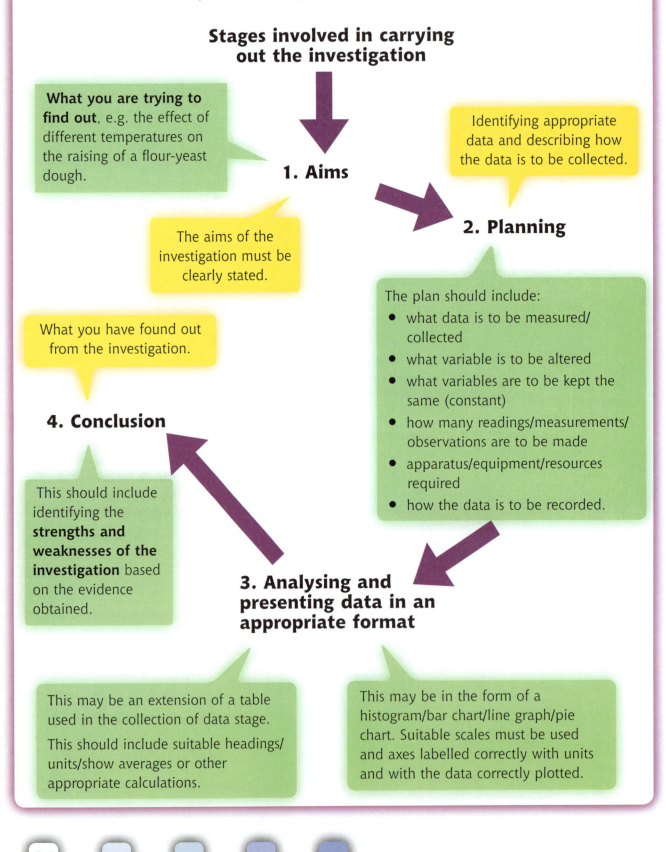

Stages involved in carrying out the investigation

What you are trying to find out, e.g. the effect of different temperatures on the raising of a flour-yeast dough.

1. Aims

Identifying appropriate data and describing how the data is to be collected.

2. Planning

The aims of the investigation must be clearly stated.

The plan should include:
- what data is to be measured/collected
- what variable is to be altered
- what variables are to be kept the same (constant)
- how many readings/measurements/observations are to be made
- apparatus/equipment/resources required
- how the data is to be recorded.

What you have found out from the investigation.

4. Conclusion

This should include identifying the **strengths and weaknesses of the investigation** based on the evidence obtained.

3. Analysing and presenting data in an appropriate format

This may be an extension of a table used in the collection of data stage.
This should include suitable headings/units/show averages or other appropriate calculations.

This may be in the form of a histogram/bar chart/line graph/pie chart. Suitable scales must be used and axes labelled correctly with units and with the data correctly plotted.

An Examination Question on an Investigation

This is an example of the type of question that you will be asked in the final examination.

(a) Yeast is used in bread making.
Name **two** other manufacturing processes that depend on the activities of yeast.

> This question can only be answered if you have learned your course notes

> Knowledge question

Manufacturing process 1 _____

Manufacturing process 2 _____

(b) An investigation into the effect of yeast on bread dough was carried out. Dried yeast was mixed with flour, sugar and water to make a dough. The dough was shaped to fit a measuring cylinder as shown in the diagram below.

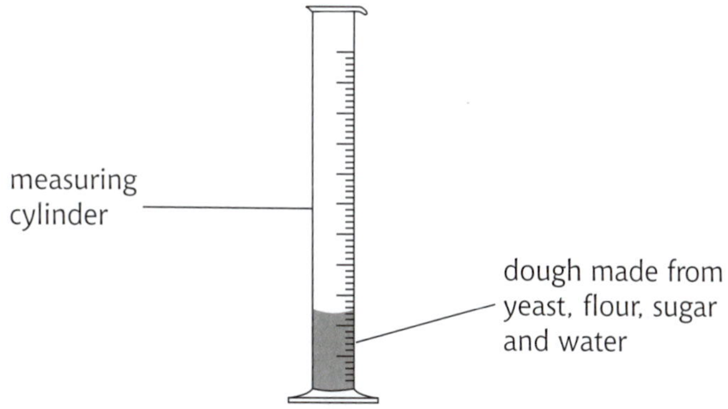

measuring cylinder

dough made from yeast, flour, sugar and water

The volume of dough was measured over a 40 minute period.

The results are shown in the table below.

Time (minutes)	0	10	20	30	40
Volume of dough (cm³)	25	27	31	37	40

(i) During which 10 minute period was there the greatest increase in the volume of the dough?

Between _____ and _____ minutes

> This question asks you to select information from the table.

(ii) Provide the axis label and scale on the grid below

(iii) Use the results in the table to plot a line graph on the grid below

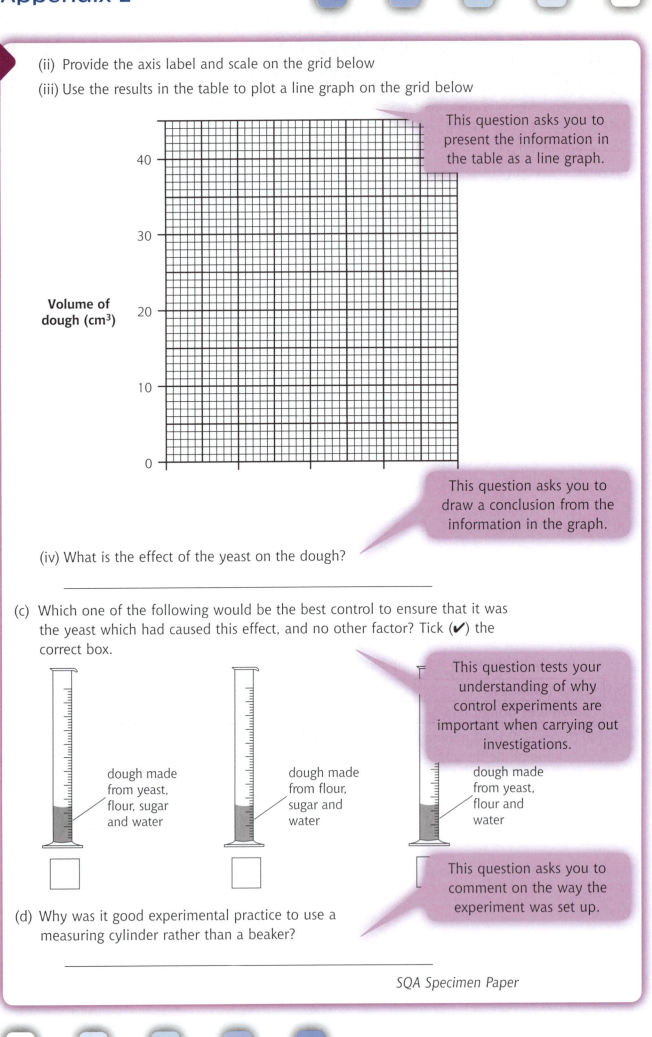

This question asks you to present the information in the table as a line graph.

Volume of dough (cm³)

40

30

20

10

0

This question asks you to draw a conclusion from the information in the graph.

(iv) What is the effect of the yeast on the dough?

_____

(c) Which one of the following would be the best control to ensure that it was the yeast which had caused this effect, and no other factor? Tick (✔) the correct box.

This question tests your understanding of why control experiments are important when carrying out investigations.

dough made from yeast, flour, sugar and water

dough made from flour, sugar and water

dough made from yeast, flour and water

This question asks you to comment on the way the experiment was set up.

(d) Why was it good experimental practice to use a measuring cylinder rather than a beaker?

_____

SQA Specimen Paper

Appendix 3 – Calculations

During the course work and in assessments and examination questions **you will be expected to be able to carry out some simple calculations**. These calculations will include **percentages, averages** and **ratios**.

Percentages

The percentage is the number or amount in a group of one hundred, e.g. pupils in a class who have a particular characteristic.

Example 1

There are 20 pupils in an Intermediate 1 Biology class.

Five members of the class are left-handed and fifteen are right-handed.

Calculate the percentage in the class who are left-handed.

To calculate the percentage who are left-handed, use the following equation:

$$\frac{\text{Number of pupils who are left-handed}}{\text{Total number of pupils in the class}} \times \frac{100}{1} = \frac{5}{20} \times \frac{100}{1} = \frac{500}{20} = 25\%$$

Example 2

(a) The table below shows the percentage of men in different age groups who are light, medium and heavy smokers.

> This percentage of men (between 16–24 years) who smoke is 13 + 19 + 6 = 38%.

Percentage of men

	16–24 years	25–34 years	35–44 years	45–54 years	55–64 years	65–74 years
Light smokers	13	7	6	3	2	4
Medium smokers	19	17	11	13	13	8
Heavy smokers	6	15	18	17	17	8

(i) What percentage of men aged 45–54 years are heavy smokers?

_____   > 17%

(ii) Calculate the percentage of men aged 16–24 years who do not smoke.

_____

SQA 2004 Paper

> So, the percentage of men (between 16–24 years) who don't smoke is 100 − 38 = 62%.

Averages

Example 1

An investigation was carried out to find out how long it takes a sample of rennet to curdle milk. The investigation was repeated five times and the times taken for the milk to curdle are recorded in the table below.

> 3. Divide the total by the number of procedures (5 in this example).

Procedure number	Time taken for clots to form (seconds)
1	37
2	24
3	44
4	31
5	29
Average time (seconds)	

> 1. Record all the numbers required to calculate the average in a table.

> 2. Add all the numbers together to get a total.

The average time for clotting to occur was

A 33 seconds — Correct answer

B 35 seconds

C 165 seconds

D 175 seconds
   SQA Specimen Paper

$$\text{Total} = 37 + 24 + 44 + 31 + 29 = 165$$

$$\text{Average} = \frac{\text{Total}}{\text{Number of procedures}} = \frac{165}{5} = 33$$

Example 2

(a) A student used a skinfold calliper to measure the skin thickness in two body areas.
   The results are shown in the table below.

| Body area | Skin thickness (mm) | | | |
	First measurement	Second measurement	Third measurement	Average
Back, below shoulder blade	8	9	7	
Front of upper arm	10	14	9	

(i) Complete the table by inserting the average measurements.   *SQA 2003 Paper*

$$\text{Average 1} = (8+9+7) \text{ divided by } 3 = \frac{24}{3} = 8.$$

$$\text{Average 2} = (10+14+9) \text{ divided by } 3 = \frac{33}{3} = 11$$

> Total

> Number of measurements

Ratios

A ratio is the **number of one thing relative to another.**

The numbers in the ratio are **always whole numbers, i.e. no decimal points or fractions**.

Example 1

There are 20 pupils in an Intermediate 1 Biology class.

Fifteen members of the class are right-handed and five are left-handed.

Calculate the ratio of the number of right-handed to left-handed pupils in the class.

Number of right-handed : Number of left-handed

15 : 5

3 : 1

> 1. Write down the numbers side by side.

> 2. Does the smaller of the two numbers divide into itself and the larger number to give whole numbers?

The ratio of right-handed to left-handed pupils in the class is therefore:

3 : 1

Example 2

A survey was carried out in a group of 45 pregnant women to find out how many of the women had smoked during their pregnancy.

Six of the women admitted to having smoked at some time during their pregnancy.

Calculate the ratio of the number of women who had smoked to the number of women who had not smoked during their pregnancy.

Number of women : Number of women who
who had smoked   had not smoked

6 : 39

2 : 13

> 1. Write down the numbers side by side.

> 2. Try dividing both numbers by 2, then 3, then 5, etc **until you get whole numbers that cannot be divided any more to give whole numbers.**

> 3. In this example, dividing by 3 gives **two whole numbers that cannot be divided any more to give whole numbers.**

The ratio of smokers to non-smokers in the group of pregnant women is therefore:

2 : 13

Answers

Answers to Appendix 1

Problem Solving and Practical Abilities

Example 1

(a)

Type of Cyclamen	Time to germinate (days)	
	Pots	Capillary matting
C. balearicum	89	120
C. coum	51	55
C. libanoticum	57	62

(b)   C. balearicum

(c)   Any one of:
   Plant the same number of seeds in pots and on the capillary matting
   Same volume of water
   Same temperature
   Planted at same time/same day
   Same spacing/area of planting
   Same type of compost in pots

Example 2

(a)   (i)   17%

   (ii)   62%

Example 3

(a)
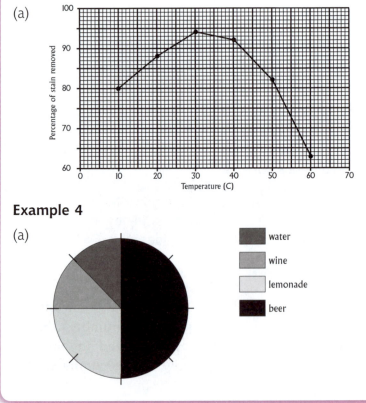

Example 4

(a)

Example 5

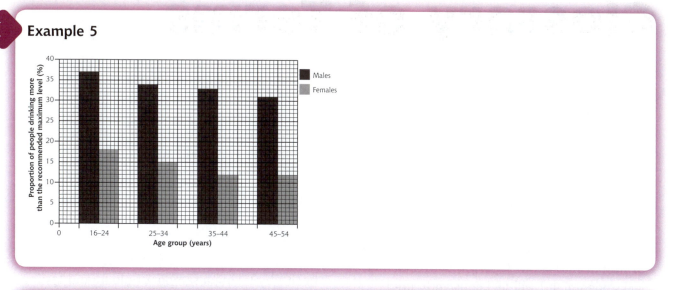

Answers to Appendix 2

An Examination Question on an Investigation

(a) beer/wine making/brewing
flavourings, e.g. marmite

(b) (i) Between 20 and 30 minutes

(ii) & (iii)

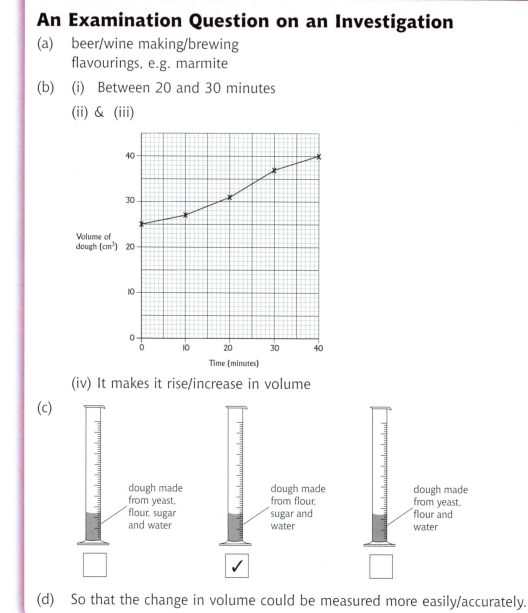

(iv) It makes it rise/increase in volume

(c)

dough made from yeast, flour, sugar and water

dough made from flour, sugar and water ✓

dough made from yeast, flour and water

(d) So that the change in volume could be measured more easily/accurately.

Glossary of Terms

Air sacs   Moist surfaces in the lungs where gas exchange takes place.

Alcohol   A chemical produced when yeast ferments sugars during the brewing process.

Algae   Microscopic plants found in lochs and rivers.

Anaemia   A condition indicated by a low level of iron in the blood. A person with anaemia lacks energy and will feel constantly tired.

Angina   A condition in which a person experiences chest pains due to the heart muscle not getting enough blood.

Anorexia   An eating disorder. A person suffering from anorexia will be severely underweight.

Antibiotics   Chemicals that prevent (inhibit) the growth of bacteria.

Antibiotic resistance   Some bacteria can become resistant to antibiotics. This means that the antibiotic is no longer able to kill the bacteria causing the infection.

Antifungals   Chemicals that are used to treat infections caused by fungi.

Aphid (Greenfly)   An example of a common plant pest. Aphids damage plants by feeding on the sugary sap in the stems and leaves.

Arteries   Thick walled blood vessels that carry blood away from the heart towards all the body tissues and organs.

Artificial propagation   Ways in which plant growers produce new plants, in a way in which plants themselves are unable to carry out, e.g. stem cuttings, leaf cuttings.

Athlete's foot   An infection of the feet caused by a fungus. It can be treated by using an antifungal cream or powder.

Balanced diet   A healthy diet that contains the main food groups: carbohydrates, fats, proteins, vitamins and minerals. A balanced diet contains several portions of each food group each day.

Biological control   This occurs when one type of animal controls the numbers of another type of animal e.g. cats kill and therefore control the number of mice.

Biological washing powder   A detergent that contains enzymes. The enzymes digest (break down) the stains.

Blood tests   A method of checking a person's health by taking a sample of blood and testing for numbers of different types of cells and levels of antibodies and other chemicals.

Blood groups   Blood exists as four main types A, B, AB and O. Blood can also be Rhesus positive or Rhesus negative.

Blood transfusion   The transfer of blood from one person to another. This may be necessary if the patient has lost a lot of blood during an operation or as a result of an accident.

Breathalyser   An instrument used by the police to measure the level of alcohol in breathed out air.

Breathing cycle   Each breathing cycle involves inspiration (breathing in) and expiration (breathing out).

Breathing rate   This is the number of breaths per minute.

Brewery-conditioned beer   Yeast is removed after the initial fermentation and any extra carbon dioxide is pumped into the beer under pressure. The beer is clear, bright like lager and has a longer shelf life than cask-conditioned beer.

Bronchiole tubes   Smaller branches of the bronchus tubes which lead to the air sacs.

Bronchus tubes (Bronchi)   Branches of the windpipe that take air towards each lung.

Bulbs   Natural vegetative plant structures that have side buds that develop into daughter bulbs.

Capillaries   Thin-walled blood vessels which allow nutrients, oxygen and waste substances to be exchanged with the cells of the body tissues.

Capillary matting   A method of watering plants using a matting that sucks up water from a reservoir and stays moist. Plant pots are placed on the matting and absorb water from it. This is a method of watering plants from below.

Carbohydrates   Compounds which provide the body with energy.

Carbon monoxide   A gas in cigarette smoke that reduces the ability of the blood to carry oxygen around the body.

Carbon dioxide   A gas that we breathe out. A gas produced during fermentation that causes dough to rise.

Cask-conditioned beer   A beer produced by allowing fermentation and carbon dioxide production to continue in the cask. The beer (ale) is dark in colour and highly flavoured. It does not last as long as brewery beer, i.e. it has a shorter shelf life.

Chitting   A method to enable seeds to germinate more easily/quickly. The seeds may be cracked or a small cut is made in the outer coat to allow water to enter.

Clinical thermometer   Instrument for measuring body temperature.

Cloche   Small polythene tunnels used to protect plants from wind, rain, frost and pests.

Compost   A plant growth medium.

Curds   Solid lumps of milk protein which occur during the manufacture of cheese.

Cutting   A small leafy side branch of a plant that is cut off and planted so that roots will grow from the cut stem. Sometimes the leading shoot is taken to encourage bushy growth below.

Dead heading   The removal of dead flowers which encourages the plant to continue to flower.

Detergent   A substance that cleans. It includes soaps, washing powders, washing up liquids, shampoos etc.

Diabetes   A condition that is indicated by a high sugar level in the blood and the presence of sugar in the urine.

Dormancy   The delay in germination of a seed until spring.

Dynamometer   An instrument that is used to measure muscle strength.

Energy balance   The body is said to be in energy balance when the energy gained from food is equal to the energy used up during bodily activities and functions. When in energy balance the body neither gains nor loses weight.

Enzymes   In biological washing powders, enzymes digest (break down) stains. Different enzymes are required to break down different types of stains. They work best at low temperatures (40°C). Enzymes are destroyed at high temperatures.

Evaporated milk   Milk that has been heated to remove some liquid making it more concentrated and creamy.

Expiration   The process of breathing out.

Fats   Provide the body with energy and insulation.

Fermenter   A container in which bacteria are used to produce large amounts of antibiotics. Fermenters are also used during the manufacture of alcoholic drinks.

Fertilisers   Substances that increase the level of plant nutrients in a compost. They are also spread over fields.

Fertiliser ratio   The ratio (relative proportions) of nitrogen (N) : phosphorus (P) : potassium (K) in a bag of fertiliser, in the order N, P, K.

Food flavourings   Substances produced from fresh yeast cake used to improve the taste and smell (flavour) of food.

Food colouring   An example of this is a pink dye added to yeast cells during the manufacture of fish food. Fish, e.g. salmon and trout in fish farms, are fed with this food causing their flesh to become pink and more attractive to the consumer.

Fungi   (singular = **fungus**) include organisms such as yeast, toadstools and mushrooms.

Glossary

Fungicide   A chemical that kills fungi that damage plants.

Germination   The development of the embryo root and shoot into the young plant. The conditions necessary for germination are moisture, warmth and oxygen.

Grey mould   A common fungal disease of plants. It occurs when plants are in very damp environments where the ventilation is poor.

Health triangle   A diagram that illustrates the three main aspects of health, i.e. physical, mental and social.

Health – physical   Good physical health involves eating a healthy diet, taking regular exercise and avoiding unnecessary health risks, e.g. smoking, drinking too much alcohol, taking drugs.

Health – mental   Good mental health involves making sure that time is found to relax and asking for help if worry, stress, low self-esteem or lack of self-confidence becomes a problem.

Health – social   Good social health involves communicating well with others and enjoying activities with family and friends.

Heart   A muscular pump in the body that keeps blood flowing through the blood vessels.

Heatstroke   A condition that occurs when the body temperature rises well above normal. It may occur when the person is in very hot climatic conditions. The person may start having fits or go unconscious or die.

High tech instruments   Instruments that are usually expensive to make and often have high operating costs. They give faster, more accurate readings and can be connected to computers.

Humidity   The amount of moisture in the air.

Hypothermia   A condition that results from prolonged exposure to very cold environmental conditions. The body temperature drops below 35°C and the person may become unconscious and eventually die if the body temperature drops to 30°C. Very small babies and elderly people are particularly at risk from hypothermia.

Immobilised yeast   A technique in which yeast cells are stuck (immobilised) to the surface of jelly beads. This means that after the yeast cells have done their job of fermenting sugars into alcohol, they can be separated easily from the product and used again.

Insecticide   A chemical that kills insect pests that damage plants.

Inspiration   The process of breathing in.

Invertebrates   Animals without back-bones, e.g. insects, slugs, snails.

Kefir   A creamy alcoholic milk drink produced in countries where the climate is very warm and people do not have refrigerators.

Lactose   The sugar found in milk.

Lactic acid   Bacteria convert lactose into lactic acid when milk goes sour. It is the lactic acid that causes the milk proteins to clot.

Layering   Involves encouraging a stem or shoot to form roots while still attached to the parent plant. The stem is pegged down in contact with the soil until roots develop. Some plants do this naturally when their stems touch the ground.

Leukaemia   A disease that is indicated by a very high level of white blood cells.

Liquid crystal thermometer   A type of thermometer in the form of a strip that changes colour with changes in temperature. It can be used to measure body surface temperature.

Loam-based compost   A compost that contains soil as one of its components.

Loamless compost   A compost that contains no soil.

Low tech instruments   Instruments that are usually inexpensive to make and use. Human errors occur more frequently when using them and they cannot usually be connected to computers.

Lungs   The organs in the chest where exchange of gases takes place. Oxygen enters the blood from the air sacs in the lungs. Carbon dioxide passes from the blood into the air sacs and is breathed out.

Maximum/minimum thermometer   Used in greenhouses to record temperatures.

Microbes   Very small organisms that can only be seen using a powerful microscope, e.g. yeasts, bacteria and viruses.

Minerals   Along with vitamins protect the body from deficiency diseases.

Muscles   Organs that are attached to a bony skeleton. When muscles contract they enable us to move our limbs.

Muscle fatigue   Occurs when insufficient (not enough) oxygen reaches the muscle tissues. The muscles become fatigued (tired) and don't function properly.

Nitrogen   A plant mineral important for good leaf growth.

Node   A point on the stem of a plant from which leaves or a side branch grows.

Non-biological washing powder   A washing powder that does not contain enzymes.

Obese   A person is described as obese if they are more than 15% above ideal body weight.

Offsets   Small plantlets produced as side shoots at the base of the parent plant.

Pasteurised milk   Milk that has been heated to 72°C for 15 seconds and then quickly cooled killing most harmful bacteria.

Peak flow   The maximum rate at which air can be forced from the lungs.

Peak flow meter   An instrument used to measure peak flow rate.

Peat   A substance that improves the water-holding capacity of a compost.

Pelleted seeds   Small seeds that are surrounded by a small ball of clay. This makes it easier to space them out more evenly when sowing.

Penicillin   An antibiotic that is used to treat a wide range of infections caused by bacteria, e.g. sore throats, chest infections, infected wounds.

Pesticide   A chemical that kills invertebrate pests that damage plants.

Phosphates   Chemicals found in detergents. They can cause increased growth of algae (microscopic plants) found in lochs and rivers. They are also a component of plant fertilisers. Phosphorus is important for good root growth.

Photosynthesis   A process by which green plants make food in their leaves in the presence of light energy. During the process light energy is changed into chemical energy.

Physiological measurements   Measurements of conditions in the body such as heart rate, temperature and blood pressure.

Plantlets   Miniature plants attached to the parent plant, e.g. leaf plantlets attached to the leaves of the Mexican Hat plant.

Plasma   The liquid part of the blood consisting of water containing dissolved substances, e.g. sugar, salts, antibodies. The blood cells are carried around the body by the plasma.

Polythene tunnels   Used to protect plants from wind, rain, frost and pests. In the tunnels, seeds can be germinated earlier in the season and crops harvested earlier. A larger crop yield is also obtained due to protection from the weather and pests.

Polythene fleeces   Sheets of polythene that are placed over plant crops to protect them from wind, rain, frost and pests. They allow water to pass through and are light enough to rise with the plants as they grow.

Potassium   A plant mineral important for the growth of flowers and fruit.

Potting compost   A growth medium that contains more fertiliser than a rooting compost to meet the needs of larger, more mature plants. It contains a 3:1 ratio of peat to sand.

Potting on   The transfer of a root bound plant to a larger pot.

Pricking out   The removal of seeds from a tray and planting them further apart to provide less crowded growing conditions.

Propagator   A thermostatically controlled container in which seeds are germinated.

Proteins   Used by the body for growth and repair of cells and tissues.

Pulsometer   An instrument used to measure pulse rate.

Reaction time   The time taken to respond (react) to a stimulus.

Recovery time   The time taken for the pulse rate to return to the normal resting level after exercise.

Glossary

Red blood cells   Cells in the blood that carry oxygen from the lungs to the body tissues.

Rennet   A substance, used during the manufacture of cheese, that is added to pasteurised milk to cause the milk protein to clot. It is obtained from calves stomachs and genetically modified (GM) fungi in fermenters.

Resazurin Test   A test carried out on milk to test its quality and suitability for drinking.

Rooting compost   A growth medium that contains less fertiliser than a potting compost. It contains a 1:1 ratio of peat to sand. It contains more sand than other composts and this allows better drainage and prevents cuttings becoming water-logged and rotting.

Rooting powder   A powder into which cuttings are dipped to speed up root development.

Runners   Horizontal stems that grow out from a parent plant and which have plantlets at intervals along their length. Strawberry and Spider plants reproduce vegetatively by producing runners.

Semi-skimmed milk   Milk that has been treated to remove some of its fat content.

Sharp sand   A gritty component of composts that improves aeration and drainage.

Skimmed milk   Milk that has been treated to remove nearly all of its fat content. There may also be some loss of vitamins.

Skinfold callipers   An instrument used to measure percentage body fat.

Sphygmomanometer   An instrument for measuring blood pressure.

Stethoscope   An instrument for measuring heart rate.

Stroke   A medical condition caused by a blood vessel in the brain bursting. High blood pressure can be a contributory factor to a person having a stroke.

Thermostatically controlled heaters   Used to control the temperature in greenhouses and keep it within the range for optimum plant growth.

Tidal air   The volume of air breathed in or out of the lungs in one normal breath.

Trickle irrigation   A method of watering plants in which water from a reservoir reaches individual pots through plastic pipes.

Tubers   Swollen roots or stems containing food that, when split, will develop into individual plants.

UHT milk   Milk that has been heated to 135–142°C for 2–5 seconds. This process kills all bacteria in the milk giving it a much longer shelf life. The treatment changes the taste of the milk.

Vegetative propagation   A method of producing plants which are identical to the parent plant and which makes use of structures such as bulbs, cuttings, runners, etc. that are naturally produced by the plant.

Veins   Blood vessels that carry blood back to the heart from the body tissues and organs.

Vital capacity   The maximum volume of air that can be breathed out in one breath after a maximum inspiration.

Vitamins   Along with minerals, protect the body against deficiency diseases, e.g. scurvy and rickets.

Water-retentive gels   Gels made of chemicals that absorb many times their own weight of water. They are often mixed with compost and used in planters and hanging baskets. This prevents these containers drying out so quickly and reduces the need for watering.

Wet and dry bulb thermometer   An instrument used to measure relative humidity.

Whey   The liquid which forms when milk clots during the manufacture of cheese. It is a waste product of the cheese making industry.

White blood cells   Cells in the blood that make chemicals called antibodies that fight infections.

Windpipe   A tube in the breathing system through which air passes from the back of the throat towards the lungs.

Yeast   A simple fungus used in bread making and the manufacture of alcoholic drinks.

Yoghurt   Produced from milk by the activity of yoghurt bacteria. The bacteria convert milk sugars into an acid which thickens the milk and gives the yoghurt its flavour. Yoghurt stays fresh for longer than milk and is a method of preserving milk products.

Index

Index